Queen
Anne's
and Lace
Other
Weeds

Queen Anne's and Lace Other Weeds

MARY JANE HARTMAN

PROVIDENCE HOUSE PUBLISHERS
Franklin, Tennessee

Scripture is based on: (1) The Living Bible. Copyright © 1971 by Tyndale House Publishers, Wheaton, Illinois 60187; (2) Revised Standard Version of the Bible. Copyright © 1946, 1952 Division of Christian Education of the National Council of the Churches of Christ in the United States of America. Used by permission; (3) Holy Bible, New International Version. Copyright © 1973, 1978, 1984 International Bible Society. Used by permission of Zondervan Bible Publishers; (4) King James Version. Copyright © 1976 by Thomas Nelson, Inc., Nashville, Tennessee; (5) New Revised Standard Version Bible, copyright 1989, by the Division of Christian Education of the National Council of the Churches of Christ in the United States of America; (6) New American Standard Bible. © The Lockman Foundation 1960, 1962, 1963, 1968, 1971, 1972, 1973, 1975, 1977 La Habra, California; (7) The New English Bible New Testament. © The Delegats of the Oxford University Press and the Syndics of the Cambridge University Press, 1961; and (8) Good News Bible. New Testament: © American Bible Society, 1966, 1971, 1976.

Printed in the United States of America

01   00   99   5   4   3   2

Library of Congress Catalog Card Number: 97–66597

ISBN: 1–57736–030–3

PROVIDENCE HOUSE PUBLISHERS
238 Seaboard Lane • Franklin, Tennessee 37067
800–321–5692

*This book is dedicated to
my two best friends:
my husband, Warren,
and my daughter, Suzanne.*

Contents

Foreword

With the volume now before you, it is my privilege to welcome you into the mind, life, and soul of Mary Jane Hartman. This collection of devotional reflections has been birthed from one unusually perceptive to the often overlooked simple messages from ordinary communicators. With a sensitivity to the everyday life around her, she has unearthed and shared valuable impressions that will alter your days.

These reflections are but a summary of the life of the author. I have known Mary Jane Hartman for thirty years, and all gathered here is but a sample of the insights and caring that is known and appreciated by her friends. During these years of friendship with her, I have read dozens (if not hundreds) of notes, letters, Christmas greetings, and devotionals reflecting her spiritual uplift from the ordinary expressions of life.

These meditations and contemplative writings have been shared for years with hundreds of persons. Many were not personally known to the author, but she knew of a need for a supporting and uplifting word. What a spiritual boost she has been to so many.

These writings come from one who knows hurt and difficulties. Throughout much of her adult life, she has struggled with physical pain preventing her from doing many of the things most of us take for granted. Even so, she has maintained a beauty of spirit and an appreciation for what she does have. Her life is a testimony to the indwelling grace of God. That which she does not have has awakened in her an unusual sensitivity to the beauty and message of the ordinary which is shared in these pages.

This volume will be interesting reading; but more, it will awaken and rekindle our own thoughts on so much that we encounter every day. Over and over, my response to these pages has been, "Yeah! . . . I have seen that. Why didn't I understand and appreciate it?"

Inside, you will discover the joy of panning for gold in some mountain stream; only this time, you will find the treasure—even treasure you didn't anticipate. Read on! Your reward for reading will far exceed your expectations.

Robert H. Spain
Bishop, United Methodist Church

Preface

My journey with this book began with a special moment.

It was one of those refreshingly beautiful spring mornings on our Tennessee hillside.

As I sat alone at our kitchen table it happened—first a word, then a thought, then phrase upon phrase filling the corners of my mind.

And, then, as though I felt the presence of a hand upon my shoulder I heard the echo of God's voice saying, "I have drawn you aside for a while. Now is the time in your life for sharing some of your life."

For some time my family and some close friends had been encouraging me to preserve some of my thoughts in written form. Perhaps now *was* the time!

There lay ahead of me many months of therapy and limited activity as I was recuperating from radical orthopedic surgery.

Thus, my therapeutic journey with the written page began. In this volume I have shared some of myself with you. These thoughts have been penned across many hours of remembrance, struggle, growth, and celebration.

Hopefully, some of these words will break in upon your life in empty and unexpected places. My hope is that they may rekindle your spirit and encourage you wherever you are on the path of your pilgrimage of faith.

Acknowledgments

I owe much of this book to those who have surrounded my life with love.

I thank God not only for tapping my shoulder, but nudging me to write through the years; for sharpening my awareness of life's lessons of value that are found in the unexpected crevices and corners around us.

Even though they did not live long enough to read much of this book, I am grateful to my parents for their special lasting gifts to me. To my father who nurtured in me an appreciation for the beautiful things in life, from a blossoming rose, to the rich patina of the wood in an antique chest, to the love of melody. To my mother who taught me the skills and instilled in me the dignity and full-blown joy of creating a home.

I recall with gratitude my denomination which entrusted me with roles of leadership over the years.

I am grateful to the special pastors and teachers who helped hone my character as I passed through the transition times in my life.

I am indebted to Andy Miller, the publisher, and to Mary Bray Wheeler, the associate publisher for product development and marketing at Providence House Publishers, who so ably and graciously helped me birth this book.

Finally, to my husband, Warren, who always encouraged me to reach for the stars and who has enabled me to catch some of them. And to my daughter, Suzanne, who has continued to give to me a lifetime of love, joy, and inspiration.

Queen Anne's and Lace Other Weeds

Think—Thank

God's Surprises

Springtime brings joy, unexpected joy! This is serendipity time in God's unfolding cycle of the seasons. Winter has long been with us, and how eagerly we watch and wait for those first hints that the earth is awakening. This is the season of expectancy, of standing on tiptoe, of awaiting the breathtaking panorama of color and form that we know will unfold before us.

Memory is a wonderful gift. But even it cannot completely store up for a whole year the glorious sights of springtime.

Perhaps my spirit of expectancy has been sharpened in recent years as springtime comes so early in our Tennessee hills, compared to those years when I lived in the upper Midwest where the springtime thaws sometimes seemed so long coming.

Once spring comes, it seems that nothing can hold it back. First comes the color of the courageous crocus as they emerge from the dead earth. Then, the brilliant yellows of the daffodils and forsythia are followed by the coral shades of japonica and the flowering hawthorne.

And then, the zenith of beauty for me comes as the first faint blushes of purple begin to show on our hillsides, foretelling that the many redbuds are about to open. Out of my kitchen window I watch those graceful arms of the redbud reach out to encourage the dogwoods close by to open their tight green buds until at last they are large and snowy white. It is then, as the dogwoods form their clouds of filmy white everywhere, that springtime comes for me.

Thank you, God, for the springtimes in your world and in our lives. Help us to be aware of them that we might use the springtime experiences in our daily lives to draw us closer to you. And amen!

Wings

Today is a gray, bleak day when one can feel spring wrestling with the chains of winter which have held the earth in a viselike grip. The many colorful birds outside my window are frantically flitting from feeder to feeder in response to the hunger signals from their tiny built-in radar systems.

And then, after satisfying this drive to survive, they fly off into the freedom of the woods. Even as they feed, they know they have wings—wings that can lift them quickly to safety when they hear the stealthy steps of the neighborhood cat or the footfall of a person on the path. It is then that they rise as a single cloud to the sky.

Just now, as I write, one little purple finch has dashed himself with a pounding thud against the windowpane. He had veered off the feeder in the wrong direction and now lies seemingly lifeless in the bed of ivy, almost at my feet. As I sit and watch him panting, I find myself wanting somehow to help him, but knowing I cannot. He is one of the lucky ones, for after several minutes, I thrill to see him stir, then stagger to his feet, and then finally spread his wings and lift himself skyward. He is wounded, but still able to fly.

He is God's creation. So am I. Many days I have watched these bits of feathered loveliness and have envied them their freedom from problems, and even have envied them their wings that lift them high above the horizon.

Today I am thinking about them again and a new understanding comes—we are much alike. God has provided for our many kinds of physical needs in different ways, expecting us to help ourselves.

Each one of us has experiences that dash us, winded and wounded to the ground, but with God's hand of courage to grasp, we can pick ourselves up and reach for the heights of life again.

God *has* given us wings—wings of vision, wings of hope, wings of faith, wings of courage, wings of love. We, too, are always surrounded by His love and care.

I am grateful for Your love and care for the birds, dear Father. I am grateful for Your love and care for me. Thank you for both of them.

We Are Connected

My heart is bursting with joy this morning as I observe the intricacy of God's creation. A mockingbird is singing in the sunshine just outside my back door. And I have picked a beautiful white peony.

Having grown up in the Midwest where peonies grow in such abundance, I have picked many beautiful bouquets of peonies. But this morning I was made freshly aware of the interrelatedness God has built into the natural world.

All of the tight buds were covered with ants. The ants were feeding on the rich, sticky substance which always covers the buds. Through this activity, the ants enable each tight bud to open into a full, beautiful blossom. How beautiful—an ant and a peony helping each other.

God has built this kind of relationship into the natural world. He has planned for it in the lives of persons, too. We need each other! We need to help each other!

The great German writer, Goethe, has written:

The world is so empty if one thinks only of mountains, rivers and cities; but to know someone here and there who thinks and feels with us, and who, though distant, is close to us in Spirit, this makes the earth for us an inhabited garden.

Another keen observer has reminded us that, "When we try to pick out anything by itself, we find it hitched to everything else in the Universe."

Life has no meaning devoid of relationships:

- wife to husband;
- child to parent;
- friend to friend;
- person to God.

Even our relationship with God first came about because of our relationship to another:

- mother to child;
- Sunday School teacher to youth;
- pastor to adult.

We learn to know God and His love through other persons. God's love for us is reflected in daily living through those who touch us.

In a recent worship service, we read these words together which affirm our need for relationships:

> We find God's presence of love with us in the Holy Spirit which often speaks to us in the voices of other people. Therefore, we affirm that openness to God and openness to others are two sides of the same coin and that really listening to others can help us hear God.

What a wise God created you and me. In His plan for our lives, He has given us precious, fulfilling relationships with others. These persons become windows which help us see Him more clearly.

Dear God: Thank you for giving us other persons to make our lives a veritable garden. Amen.

Standing on the Promises

On the occasion of her birthday, I bought a little book to give to my granddaughter as a birthday present. The title of the book was *Standing on the Promises*. As I thought about the title, I became aware of the many promises from His Word, that I regularly and repeatedly claim for myself. Here are some of my favorites:

My times are in thy hand.
O God, I wish them there.
My life, my friends, my soul
I leave entirely to Thy care.

Because the Lord is my Shepherd, I have everything I need!
—Psalm 23:1 (LB)

Thou wilt keep *him* in perfect peace, *whose* mind is stayed *on thee.* . . .
—Isaiah 26:3 (KJV)

The Lord is my light and my salvation; Whom shall I fear?
—Psalm 27:1 (NASB)

Fear not, for I am with you. . . . I will help you; I will uphold you with my victorious right hand.
—Isaiah 41:10 (LB)

Don't be anxious about tomorrow, for God will take care of tomorrow, too.
—Matthew 6:34 (LB)

Help me to rest in the hours of today and entrust my future to you.

We are not at our best when we are perched on the summit. We are at our best when we are climbing—even when the way is steep.

Mother Teresa has given us this humorous quote: "I know God will not give me anything today I can't handle, but sometimes I just wish He didn't trust me so much."

Underneath are God's everlasting arms. But we will only feel them and gain strength and support from them when we lean back in His arms in faith and trust.

What the caterpillar believes is the end of the world, God turns into a butterfly.

The Lord is near. Do not worry about anything, but in everything by prayer and supplication with thanksgiving let your requests be made known to God. And the peace of God, which surpasses all understanding, will guard your hearts and your minds in Christ Jesus.

—Philippians 4:5–7 (NRSV)

Dear Lord: We thank you for the wonderful promises you have given us which enrich our lives. We claim them now, in faith believing. Amen.

Love and Miracles

The late Willa Cather has left us a powerful and commanding thought: "Where there is great love, there are always miracles."

Does great love always generate miracles? Webster defines *miracle* as "an event or action that contradicts *known* scientific laws," or as "a remarkable thing."

In the first-century Christian era, the Greeks used six different words to convey the different meanings of the word "love":

1. Eros: the love of a man or woman for a person of the opposite sex;

2. Phileo: the social love or affection for a friend;

3. Stergo: the love and affection among all members of a family;

4. Philadelphia: the love between siblings a.k.a. Ben Franklin's city of brotherly love—Philadelphia;

5. Philanthropia: the love of persons or organizations who give to help humanity;

6. Agape: the unconditional love of God for us—His children.

However, the New Testament writers used only two of those six Greek words when they wrote about love: *phileo*—the love of friends and *agape*—God's love for us.

In the New Testament, phileo was used to describe the love of the early Christians for each other. Their love was so deep, powerful, and real that it caused others to say, "See how these Christians love one another." Their love for one another and for Christ fueled the miracle of courage in their lives.

Later, the manner with which the catacomb Christians bravely endured the cruel inhuman persecutions was a powerful witness to the whole world. Repeatedly, down through the centuries since that time, Christians have banded together as they courageously and compassionately witnessed to their common faith.

The second word for love which the New Testament writers used is agape. This is the highest form of love, freely given with no thought of return. God loves us, regardless. It was this kind of love about which St. Augustine wrote when he said, "God loves each one of us as if there was only *one of us to love.*"

We see a form of this kind of love when a parent loves a child unconditionally, even though the child has strayed. This kind of abiding love is not diminished between friends, even though one has been deeply hurt by the other or disappointed in the other. Agape love always plants seeds for the miracle of a healing reconciliation.

Even more importantly, we can know the greatest expression of God's agape love through Jesus Christ, our Savior. Whenever His hand touches our lives, He leaves fingerprints of love *and* miracles.

Refreshment

Thank you, God, for this slow, gentle rain outside my window. Even now, I can smell the dry, parched earth as it thirstily drinks in the drops. I can see the brown blades of grass being washed clean again. Surely, they are beginning to lift their lifeless spines as they absorb the wet refreshment.

The leaves on the trees, which have curled up from the scorching sun, are hesitantly unfolding their edges to the heavens. Oh, how our earth needs this life-giving renewal after so many arid weeks without your refreshing rain.

So, too, my soul needs to be showered to know the soothing relief from the parched, dry hours it has known. It is so easy to feel fresh and fruitful when the rains of the good days fall. But where is the reservoir in my heart I thought I had filled for the dry, arid times?

My being was refreshed today through the heart-to-heart sharing with the daughter You gave us years ago. Like "springs in the desert," her needs and hopes helped bring new life and purpose to mine.

How we all need each other to help fulfill our dreams; child to parent, friend to friend, student to teacher. We know these persons, and their relationships come from You, Father.

Thank you for the healing refreshment that you send through those about us whom we love the most. Thank you for this slow, gentle rain outside my window.

When Your Heart Says Thank You

Human happiness depends upon our ability to be grateful. In one of his orations, Cicero said, "The grateful heart is not only the greatest virtue but the parent of all others." The person who has a grateful heart has almost everything, whereas the person whose heart is ungrateful has almost nothing.

And so it is with our happiness. It is found in the small everyday things and places we stumble over blindly. It is everywhere about us.

It is appropriate for us to think of thankfulness. Happiness and thankfulness are so closely bound to one another. Really thankful persons are happy persons and truly happy persons are always thankful ones. Just as we may be prone to look for happiness over the next hill or during the next day, so we must train ourselves to look for it in this very moment and in things close about us.

Over the doorway of a small chapel in England are the two words: "THINK—THANK." If we will only stop to *think*, we will stop to *thank*!

Luke tells us the story of the ten lepers whom Jesus graciously healed. We recall that only one returned to say thank you. How many times in our normal days do we pause to say thank you to God for our many daily blessings?

If we would attempt to make a list of things for which we should give thanks, it surely would be an unending one. May I share a few from my long list?

Today I am so thankful for those "golden gems" from God whom we call friends. Years ago, Dr. William Stidger wrote a thank-you letter to a former aged school teacher to say thank you to her for all she had meant to him. In her reply, she said that she had been teaching school for fifty years, and his was the first letter of appreciation she had received. Since reading this, I have tried to send one letter of thanks to such a person each year at Thanksgiving.

Next on my list of thanks are those gifts I call "gateways to God." The privilege of worship, the enrichment of Bible study, the open line of prayer, the joy of sacrificial giving and service. Through these spiritual disciplines, we enter into the very presence of God Himself.

The next consideration is the most difficult for me. It is when I struggle to give thanks for the hard things that life brings to me.

Into the city of Eilenburg, Saxony, during the Thirty Years' War, refugees came by the thousands. Soon they exhausted the food supply, causing a famine. Later, a pestilence broke out and before long many were dying. The only surviving clergyman in the city, the Rev. Martin Rinkart, tried to comfort the sick and dying. On some days he conducted as many as fifty funeral services. And it was on just such a day that he sat down at eventide, weary and exhausted, and penned those wonderful words of thanksgiving which we sing during the Thanksgiving season:

Now thank we all our God
With heart and hands and voices,
Who wondrous things hath done
In whom His world rejoices;
Who, from our Mother's arms
Hath blessed us on our way
With countless gifts of love,
And still is ours today.

—Martin Rinkart

We need to give thanks for times of trouble and hardship.

Of course, the epitome of our list is reached when we are thankful to Christ and the salvation He offers to us. Surely, this is our prime reason for giving thanks. When we think of God's greatest gift to us, our hearts must respond in the most profound ways we know.

Were thanks with every gift expressed
Each day would be Thanksgiving!
Were gratitude its very best
Each life would be Thanksliving!

—Chauncey R. Piety

A Rose in December

Some months ago a friend sent a very special greeting card to me. On the front of the card was a beautiful bicolored rose and these few lines of verse:

God gave us memories
So that we might have
Roses in December.

I have pondered the thought in these words over and over again. I have come to realize that God's precious gift of memory enables us to remember:

- a glorious sunset even after we can no longer see it;
- the laughter of our children even after they have become adults;
- the caring of a friend even after she has moved miles away;
- the touch of a loved one even after he is no longer here; and
- a rose in December even though it bloomed in June.

Along with this special gift of memory God added another bonus—our ability to cushion the painful hurt of some remembrances and to focus on the happiness our memories bring.

Burton Hillis, writer of "The Man Next Door" in the *Better Homes and Gardens* magazine, adds his light touch to this thought. "Nostalgia is a powerful force when it can convert the days of the Depression into the good old days!"

God's greatest healer is time, and this is the cushion He gives us to soften many of our remembrances:

A mother tends to forget the pain of childbearing and remembers the joy of the newborn baby in her arms.

Parents tend to forget the pain of a daughter losing an eye and remember the strength of her vibrant faith.

A man tends to forget the pain of losing his job and remembers the flood of support from his friends.

A daughter tends to forget the pain of her father's death and remembers the deep security of his caring.

Yes, God cushions the hurt and magnifies the joys of remembrance. They are like treasured Christmas gifts God holds for us to be opened again and again. I thank God for memories that bring roses in December.

Handle with Prayer

Handle with Prayer

In a recent worship service someone said, "Life is fragile. Handle it with prayer." As I look back on my life, many experiences flood over the sea of my memories. There were happy, joyous, and fruitful times. There were also difficult, painful, and desolate times. Now, I realize that I could not have possibly emerged from all of them as a whole person without the soothing, comforting, and challenging therapy of prayer.

We mortals are such impatient beings. We forget to measure things in terms of God's timetable instead of our own. Perhaps there is no area in our life where this is more apparent than in our prayer life. We hurriedly utter a prayer and almost in the next instant we expect an answer to our liking. Slowly, we must learn that God will answer every prayer, but at a time and in the manner that is right for us.

Dr. Nels Ferré, a prominent scholar of a generation ago, referred to God's "majestic yes" and God's "eloquent no." We would add that God sometimes says, "Not just yet." At other times, He says, "Get busy."

In God's infinite wisdom, He often does say, "Yes." There are times when our askings are in accordance with His will. It is important to remember that God always wants only the best for us.

At other times the answer is, "No." He sees much farther down the road than we do. There are also times when His answer is, "Not just yet." He knows that in some ways we are not ready for an affirmative answer.

And then, so many times His answer is a loud and clear, "Get busy!" One night a little girl prayed, "Dear God, thank you for not letting one of your birds get caught in Jimmy Brown's bird trap. Jimmy is a nice boy, but he does such naughty things. Amen."

Her mother, who was hearing her daughter's bedtime prayer, asked, "How can you be so sure that God will not let any of His birds get caught in Jimmy's trap?"

The little girl's confident reply was, "I know He won't, because today He helped me smash that old trap all to pieces."

Sometimes, the answer to our prayer is very different than we expect or hope for. Phillips Brooks must have had this in mind when he wrote:

> Do not pray for easy lives; pray to be stronger people.
> Do not pray for tasks equal to your power; pray for power
>   equal to your tasks.
> Then, the doing of your work shall be no miracle, but you
>   shall be a miracle.

> Every day you shall wonder at yourself, at the richness of
>   life which has come to you by the grace of God.

God does hear every prayer. God does answer every prayer. *Your* life is fragile. Handle it with prayer.

A Burning Bush

Recently, as we traveled the gray-ribboned highways of New England and the Canadian Maritime Provinces, I was thrilled anew with wonderment. For miles and miles we followed a trail dripping with color that God's paintbrush had left behind.

> From the greens of the cedars and pines to the golden
>   ambers of the sycamores;
> From the rich bronzes of the oaks to the fiery reds of the
>   sumac and maples.

All of these monuments of the mountains and hillsides seemed to close in and enfold us.

As I looked at a glorious flaming red maple, I was reminded of a burning bush on another mountainside. It changed the life of a man and the destiny of a nation.

God still sends burning bushes that can change our life's direction. We can hear God's call:

- through the printed page;
- through the lyrics of a song;
- through the beauty of His world;
- through the voice of a friend;
- through the cries of the hungry.

Are we watching? Are we listening?

It's Time to Pray

The forty days of Lent represent the forty days that Jesus spent in the wilderness. We can be sure that He spent much of that time in prayer when we recall the three extreme temptations the devil posed to Him. It was a time of concentrated preparation for His ministry in the days to come.

For the Christian, Lent is a period of introspection—a time of looking within our souls and then reaching out for strength and direction. This reaching out, we call prayer.

What is prayer? Dr. Harry Emerson Fosdick, one of the greatest preachers of our lifetime, has given us this definition of prayer: "Prayer is not begging *from* God; it is cooperation *with* God. It is simply giving the wise and good God an opportunity to do what His wisdom and love want to do." Somehow, we may say that prayer is simply being with God and knowing it.

It is important to consider the *when* and *where* of our praying. We surely have many unused prayer times in our daily schedules. Someone has said, "Though my head and my hand be at labor, yet doth my heart dwell in God." We can pray when doing those routine tasks that do not require creative thinking.

Prayer is but thought in tune with God
And often through the day
My heart becomes an altar
Where my soul kneels down to pray.

There are several parts we may want to include in our prayers:

1. Adoration, praise, and thanksgiving: "O God, I love you and thank you."

2. Confession: "I am sorry—I ask your forgiveness."

3. Petition: "I want this, but not my will but Thine be done."

4. Intercession: "Please help John and Mary."

5. Commitment: "I give myself to you, O God—use me."

6. Assurance: "In you, God, I have my hope for tomorrow."

Not every prayer will include all of these elements. A given prayer may center on only one element.

Slowly I Have Learned

Slowly I have learned God answers prayer.
Slowly I have learned this vital thing;
That my petition loosed upon the air
Will reach its destination, and will bring
The answer that will be the best for me inevitably.

Three large trees in a forest prayed that they might choose what they would become when they were cut down. One prayed to be made into a beautiful palace; the second, to be a large ship to sail the seven seas; and the third, to stay in the forest and always point to God. One day the woodsman came and chopped down the first tree, but instead of a palace, it was made into a stable, wherein was born the fairest Babe of all creation. The second tree was made into a small ship that was launched on the Sea of Galilee, on the deck of which stood a tall young man who told the multitudes: "I am come that they might have life, and have it more abundantly." The third tree was made into a cross and to it men nailed that young man, the loveliest personality who ever walked on the earth. Ever since then, that cross has been pointing men to God. And so, each prayer was answered.

Prayer can change your life. Prayer can change our world.

Dear Father: Teach us *how* to pray. Grant us faith to pray so that we deserve to be heard. Amen.

Monday Morning Prayer List

*F*am sitting at my kitchen table and have just made my "Monday Morning Prayer List." This has become as much a part of my Monday morning routine as the laundry I still like to do. That practice is a carryover from the old Monday washday custom.

This is the beginning of a new week. What better time than this to recall on paper the names of persons I want to intentionally draw into the circle of my love and concern for the week.

Sometimes, the same name will go on the list for two weeks, several months, or even longer. Often, the happy word will come that a crisis in a person's life has been met. After a prayer of thanksgiving, I will delete that name from the list. Sometimes, word comes which indicates that circumstances have changed, such as when death occurs. Then, the names of those who are grieving go on the list. Some Mondays the list will have many names, and on other Mondays just a few.

There is great power in intercessory prayer. In intercession, we unselfishly enter into God's purpose for others and become coworkers with Him. Intercession is identifying ourselves with others in their needs and in their joys.

At one time, I served as a volunteer in the Upper Room Prayer Center. Through a WATS line, any person from anywhere in the country could call the prayer center when they felt a need for prayer. Sometimes, callers were so emotionally distraught that they could not pray for themselves. A volunteer would pray with the caller, offering the hope and help that comes through prayer.

On those Tuesday afternoons as I answered the calls of those who were asking for help, several things happened to me:

When I prayed with another, I also became more aware of the hurts of still others.

When I prayed with another, my own personal problems faded in light of his or her problems and concerns.

> When I prayed with another, I gave my personal thanks that God has given us an open communication line with Him.
>
> When I prayed with another, I knew that God would give an extra surge of emotional and spiritual strength and coping ability to the caller.

There were shadows of hurt and pain in my life. My friend faithfully prayed for me when I could not pray for myself.

Then came stronger days for me and it was a joy to pray with another friend whose life had become difficult.

The answer to my prayer came gradually in bits and pieces. Her answer came in a single moment of quiet that was so intense, it seemed even the silence could be heard.

In each life, hers and mine, God's power was released to give comfort, strength, and hope. This is intercessory prayer at its best.

The Unbroken Path

The excellent resounding voice of Charles Kuralt is no longer heard on his popular Sunday morning program, *Eye on America*.

Several months ago, he described the completion of the last link in the longest walking path in America. This trail is known as the Pacific Coast National Scenic Trail. It is 2,638 miles long.

The dream of completing such a trail from Mexico to Canada began in the hearts of several nature lovers in 1920. It took seventy-three years to fulfill that dream! The trail begins on the Mexican border just east of San Diego, and ends in the Manning Provincial Park in British Columbia, Canada.

The dreamers begged, bartered, and bought the rights to finally connect the several segments of this walking trail. It is a simple, narrow path. The path affords the walker time to reflect. Here one can see the lush growth of wild flowers; nearby trees are reflected in quiet streams; and the splendor of both Mount Hood and Mount Ranier can be seen along the way. The path leads through quiet, deep canyons, across easy walking level areas, and up steep climbs to breathtaking vistas of three different mountain ranges.

The trail affords the hikers "islands of silence" to be quiet, to ponder, and to meditate. It is hard for great creative thoughts to come to one in the din of a noisy eight-lane interstate highway with cars on either side jockeying for the best position.

None of us may ever walk on even a small part of this unique and natural "getaway" trail. But each of us must find our own quiet places and times to center our thoughts and our beings everyday. Otherwise, our lives become like the chameleon placed on a plaid jacket—we explode in all directions. We are hurried and harried throughout our days. We are exhausted when night falls and we feel we have accomplished little. Surely, we must find the way to a slower pace.

The psalmist heard the words, "Be still and know that I am God."

The hymnwriter must have been thinking of this verse when he wrote, "There is a place of quiet rest, near to the heart of God."

May each of us find our own special unbroken trail which will lead us into the very presence of God.

Please Listen

O ne of God's greatest gifts to us is the gift of hearing. Someone has suggested that we do not really take the time to listen to Him or to each other. God does not ask us to *do* as much as He asks us to *listen* in order that we might have His direction.

What are some of the reasons we do not really listen to each other? Let me list a few that are problems for me:

I am so preoccupied with my own agenda—my thoughts and problems.

I am not interested in what another is saying as I have heard it before.

I am busy thinking about what I want to say next.

I notice another person close by with whom I wish to speak before they move on.

I do not feel the person who is talking is saying anything valid. . . .

The novelist, Adela Rogers St. John, once asked the actor, Gary Cooper, who his favorite actress was and why. His quick reply was, "Ingrid Bergman—she listens to me with her eyes instead of thinking what her next line will be."

In his book, *A Taste of New Wine*, Keith Miller has written these beautiful words about listening:

I have found that real contact is made with another person mostly by *listening*. What you are doing when you are listening as a Christian is putting your hand quietly in the other person's life and feeling gently along the rim of their soul until you come to a crack—some frustration, problem, or anguish—you sense that they may or may not be totally conscious of. As you listen this way, you are loving this person and accepting him or her just as they are.

There is a magical bond in this kind of listening.

Of even greater importance is our listening to God. Is our communication with God, which we call prayer, always a one-way line of asking, asking, asking, rather than listening?

Listening to God is not always easy because it requires us to:

- turn off all our impatient requests;
- turn off our personal concerns, including our children, the dishes, the telephone, the computer, the work we brought home with us; and
- shift our souls into neutral in order to tune in, to be quiet, and to listen actively.

Being quiet alongside God will enable us to better know:

- who He is;
- who we are;
- why we are here; and
- what He wants us to do with our lives.

The French priest, Michel Quoist, has written these thoughtful lines in a poem, which he calls *The Telephone*:

I have just hung up; why did he telephone?
I don't know . . . Oh, I get it. . . .
I talked a lot and listened very little.
Forgive me, Lord, it was a monologue and not a dialogue.
I explained my idea and did not get his;
Since I didn't listen, I didn't help,
Since I didn't listen, we didn't commune.

Forgive me, Lord, for we were connected
And now we are cut off. . . .

(From *Prayers by Michel Quoist*, Avon Books, a division of Hearst Corporation, 959 Eighth Avenue, New York, New York, 10019, 1975.)

Listening to God can change the quality of our caring. Listening to God can nurture our souls. Amen and Amen.

Islands of Silence

We all have a need for "islands of silence" in our lives. We constantly feel buffeted about by the pressures of everyday living. The calendar and the clock have become the lords of our lives.

In his book, *On Beginning from Within*, Douglas Steere was mindful of the pressures of our day when he wrote:

> In the urban life of the western world we live in a veritable hail of stimuli that solicit our continued response. The tempo of our lives has increased until we feel continually "driven by life."

It is very difficult to function and grow emotionally and spiritually when we are under a constant pressure to survive.

In the midst of his busy life, Henry Wadsworth Longfellow made a discovery which he describes in these words:

> The holiest of all holidays are those kept by ourselves in silence and apart; the secret anniversaries of the heart.

Henri Nouwen, one of our contemporary writers, when writing of our drive to keep busy, adds this note of caution:

> We think we are too busy to pray; we have too many needs to respond to, too many wounds to heal. We feel that prayer is a luxury, something to do during a free hour, a day away from work, or on a retreat.

I do not often find that free hour, that day away, or that retreat experience in my daily routine. But I have found some special moments in my day, some "islands of silence." Perhaps you have found them, too. These are the blocks of time when I am doing the many tasks in my home that require little thinking because they are so routine: making the bed, unloading the dishwasher, folding the laundry, ironing a blouse, or dusting the furniture.

These have become special times to me. My mind is always filled with some thoughts, and I have realized that many times those thoughts are not very important. Now, during those routine tasks, I try to empty my mind of those unimportant thoughts to make some space for God's thoughts. This has become a time of just *being*, a time of being present with God.

> Quiet now . . . close the mind's door on the business of the day and for this brief moment clear the way for God.
>
> Quiet now . . . no need for words . . . listen . . . and be still . . . His voice will direct . . . His Spirit will fill your soul.
>
> Quiet now . . . breathe in new strength, new courage. Learn His master plan for you . . . then, refreshed, return to duty.
>
> —Author Unknown

God has time to listen if we have time to pray.

Sparkling Shiny Windows

Sparkling Shiny Windows and Snowy White Curtains

*T*hey are beautiful, simply beautiful! The morning sun shines through the clean, crystal-like panes we have just washed and brings the out-of-doors indoors in clearer, sharper focus. This picture is set in the ruffled frame of freshly laundered white curtains.

Why have I let my procrastination keep me from the joy and sense of accomplishment that these clean windows and freshly laundered curtains bring me today? I have procrastinated in the past, and knowing myself as I do, I probably will again in the future.

Somehow, I will carelessly put off doing those things that bring me joy. I will thoughtlessly wait to do those things that could bring joy to others also. "Do not put off until tomorrow what you can do today." What words of wisdom we have known so long, yet have heeded so seldomly.

A sign on the desk of a prominent man dying of cancer reads, "Do It Today!" He does not know how many tomorrows he may have. You and I do not know how many tomorrows we may have and our yesterdays are gone forever.

Yet, I do hold this day, this single moment in my hand. It is my gift of "now" from God. It is really all I have to use and to give away.

49

Perhaps this is the motivation for the lists of "things to do" which I make each day. They are the simple, quiet tasks involved in everyday caring:

- making a phone call to Lois who is lonely;
- preparing a meal for Margaret who is moving from our neighborhood today;
- visiting a friend who is homebound;
- writing a note to Stella, my favorite teacher;
- baking those special muffins my family likes; or
- listening to Marjorie as she struggles through a problem.

All of these along with the routine tasks of my day make my list long. When the day is done and I cannot cross off every item, I feel frustrated. Perhaps it is because I sense the urgency of too many needs unmet, so many persons not helped, so many hopes unrealized.

My lifetime cannot give me *enough* time to do all the things I want to do. And so, I add the unfinished things from today to a fresh new list for tomorrow, just hoping that I may still have another day to enjoy these sparkling shiny windows and snowy white curtains. . . .

There Is Power in Your Influence

Some of us live alone. Some of us live in families. Some of us live in communities of families. All of us are related to others in a number of ways. One of the strongest words that describes the way we relate to others is the word *influence*.

Webster defines *influence* as "the power to affect others or the power to affect change." Our influence is like a current of power flowing into the lives of others—those with whom we come in contact everyday.

Your influence is like the little pebble that a child throws into a quiet pond. As it hits the water, its impact causes one ripple, then two, then three, and then even more as the ripples spread farther out in ever-widening circles on the surface of the water. So, the power of your life through your influence flows out and affects the lives of many others near and far. Your spouse, your children, your grandchildren, your friends, your neighbors, in fact, everyone that you meet—all are affected in some way by your influence.

Some years ago, we refinished an antique cherry cradle. As I rubbed on the beautiful wood for several hours, I wondered what had happened to all of the babies who had been lulled to sleep in that old cradle. I was reminded of the age-old adage, "The hand that rocks the cradle rules the world."

In a 1996 poll of college-age women, the Scripps Howard news service asked them to name their women role models. Surprisingly, the names of highly visible women in public life did not appear at the top of the list. Instead, the majority of those young women said they were much more inspired by their mothers, grandmothers, cousins, and aunts, than by public figures.

As parents, grandparents, uncles, aunts, teachers, and neighbors, we should never underestimate or relinquish the power of our influence within our families, our churches, and our neighborhoods. Our children, our grandchildren, and all young lives we touch are like jewels to be polished and to be presented back to the Lord. We can give no greater gift to Him and to the world than a young man or a young woman who, in turn, uses his or her influence to create a better world in which to live.

Our influence cannot be weighed, measured, or wrapped in a neat little package. It is one of the most powerful forces we have under our control.

Just for a moment, think of those persons who have greatly influenced your life. Try to remember that the kind of influence you and I spread is strongly affected by an even greater influence—the presence of Jesus Christ in us.

Who you are as a person and what your quiet influence demonstrates everyday, speaks even more loudly than what you might say to those around you.

Over My Shoulder

*D*uring these beginning days of another New Year, I am looking back over my shoulder at the prints my steps have made this past year.

I ask myself what I have really done with my life the last twelve months. How have I used those 365 days to make this a better world?

Suddenly, the thought breaks into my quiet time with a thunderous crash. Does it really matter at all that I have lived this year? If God gives me the gift of life in the new year, will my life make any difference in the lives of those I will touch? Will it really matter that I have lived on this planet, in this city, in this neighborhood, in this church, in my home?

Someone has suggested:

Man's greatest goal is that his life will make a difference in
the world.
Man's greatest joy is knowing that everyday his life *does*
make a difference in the world.

Surely, my life and yours will make a difference if we will:

- wipe away the tears of a child;
- be sensitive to the hurt of a loved one;
- give hope to the "down-and-outer" who is discouraged;
- rejoice with the achiever in his good fortune;
- reach out a hand to welcome a newcomer;
- bestow praise on one who has done well;
- put an arm around one whom death has left lonely;
- encourage some forgotten one to try just once more;
- pray for one who has lost her way in depression;
- hold the hand of another who has a life-threatening illness.

God can take our little bits of love and caring and multiply
them over and over again as they flow into the lives of others. The
deeds we have done may be forgotten. But the lives we have
touched and changed will continue to help the world after we have
left it. To live on in the lives of others whom we leave behind is not
to die.

As we reexamine our lives, we realize that we have given our
time and energy to those things that we have made our priorities.
Let us pray that we may keep our priorities in order. Let us believe
that God can and will do the unexpected *for* us, *with* us, and *through*
us in this new year.

What Really Counts

Today is your day and mine; the only day we have; the day in which we play our part. What our part may signify in the great whole, we may not understand; we are here to play it, and now is our time.

—David Starr Jordan

This is cupboard, closet, and drawer-cleaning month for me. The pace of my cleaning always slows down when I get to the file drawers beside my desk. Here I keep a folder labeled "Seed Thoughts and Quotes." And today I have spent hours reading through some of the things I have clipped and saved this past year.

The above quote that I found in the folder has given me cause to stop and take stock of my hours, my days, and the ways I spend them. You and I have a debt to the yesterdays which we can only repay today.

Someone has suggested that we must each work selflessly for the things we love, and should regard our own contributions as a link in a chain which started long ago and will reach far into the future.

Our grand business is not to see what lies in the distance, but to do what lies clearly at hand.

Emerson reminds us: "You cannot do a kindness too soon, for you never know how soon it will be too late."

The well-known historian, H. G. Wells, once put it in these words:

Subordinate and everyday things surround me in an ever-growing jungle. My hours are choked with them; my thoughts are tattered by them; the clock ticks on, the moments drip out and trickle, flow away as hours. I am tormented by a desire for achievement that overruns my capacity.

So, then, it is for us to decide what things are really important. We must sift through the rubble to find the shining and enduring shapes and patterns our lives would create and leave.

I am reminded again and again of one short sentence my grandmother wrote to me many years ago. She sent it at a time when she sensed my weariness and discouragement from my taxing involvement in the many activities of the local congregation my husband was serving:

> Remember, my dear, only what you do for the Lord really counts. Only those things you do for others in His name will endure.

That was the credo by which she lived. Surely an appropriate epitaph on her gravestone would have been, "She lived for others today."

You and I cannot do *all* things for the *whole* world; but we can do *some* things for *part* of the world nearby; and we should do them *now*.

The Monday Morning Clothesline

Several years ago, my eight-year-old granddaughter and I were browsing in an antique shop. She asked me the names and uses of many things. When she held up a square, wooden clothespin and asked, "Granny what is this?" I knew I was really dated.

I am more than old enough to have lived in the days when the clothesline with its square, wooden clothespins was the only way we had to dry our laundry.

I thoroughly enjoyed hanging up those fresh, clean-smelling clothes. The bluish white ones first and finally the dark socks hung in rows by the toes. Those square, wooden pins held them in an orderly, symmetrical pattern, each piece next to one of its own kind.

I always had a warm feeling that I was taking good care of my family when I surveyed those neatly hung pieces flapping in the gentle breeze. Many times as I looked at their varied shapes and sizes I recalled the words of my great-grandmother: "Always remember that cleanliness is next to godliness." She practiced both in her life each day for all the years I knew her.

There was another fringe benefit from those Monday morning washdays. They were the special times when my neighbor and I visited over our rose-covered fence. We caught up on the happenings in our families.

There were joys and sorrows shared. We also talked about hopes and disappointments. Underneath it all there was a deep sense of caring concern for each other. How I miss those Monday morning washdays!

Now, I live in a subdivision where we are not allowed to have a clothesline. I take my laundry out of the washer and put it into a dryer alongside the washer.

So, I must find new ways of replacing the Monday morning clothesline. Between loads I can use my telephone to check on an elderly neighbor. I can use the pen at my desk to write a note of concern to another. I can put a freshly baked pie at the back door of the home of a nearby working mother for her to give her family that evening. I can baby-sit for a young mother who is caught up in the monotony of diapers, dishes, and dirt so that she may have an afternoon out.

There are many other ways I can replace that Monday morning clothesline. God needs each of us to love and care for those who live nearby. Why not "reach out and touch someone" today? Perhaps God is counting on you to do it.

God's Nudges

believe that God expects us to help care for His children. The theologian would say that God nudges us through the activity of the Holy Spirit.

Others would describe God's nudges as that still small voice deep within us as it whispers to us.

Surely, God nudges us in the listening time of our prayers when our hearts pick up radar-like signals from His great heart.

Still others would suggest that God nudges us through a pervasive restlessness down deep within our very being.

Regardless of the way God speaks to us, He does give us definite nudges. In her book, *Christy*, Catherine Marshall relates numerous stories of the way God nudged folks in Appalachia. Christy, a sensitive young teacher, averted several tragedies in the lives of her students when she responded to God's nudges. Each of us has experienced some of God's nudges in our lives.

I vividly recall a moment during a worship service when I sensed a nudge to call a friend who needed the services of an honest and dependable lawyer to help her solve some complicated problems.

Or, on another day when I felt I must write a note of caring to a friend, not realizing it would arrive on the day she would experience a crisis.

I heard recently of a Catholic priest in California who felt God's nudge in a different direction. He went to Romania and established a loving home for HIV-positive babies who had been living in deplorable conditions as wards of the state. When a New Jersey couple heard of the project, God gave them a nudge. They went to Romania and adopted five of the children and brought them back to America. Love takes us to places where our minds would never let us go.

Then, there was the day when I felt "down" emotionally. A thoughtful friend telephoned to say she and her husband were bringing in our evening meal and would stay and eat it with us. She surely felt one of God's nudges.

I am convinced that God does nudge us to help tenderly care for His children. I pray that I might always be sensitive and obedient to those holy pushes.

Those Little Lights

As we sit on our deck this evening the fireflies are playing hide-and-seek all around us. I see them here, and then in another place. They remind me of tiny flashing neon lights—on again, off again, on again. . . .

These little lights against the darkness of the night are not large or spectacular in any way. But their unanticipated bits of glow and loveliness add something special to this warm summer night. The fireflies give what little bit of light they have and then move on to light another spot.

Many of us waste our lifetimes waiting for the big, flashy, and spectacular jobs we think God has for us to do while all the time we forget that each of us has at least a one-talent light which could glow on quiet hillsides and in the valleys of the world about us. Our lights may never be seen by many nor loudly acclaimed by the masses.

Two lines from that old chorus I learned as a child come back to me through the quiet of this evening:

This little light of mine,
I'm gonna let it shine.

Recently, I heard a stimulating and challenging sermon on the subject, "The Sin of Doing Nothing." The speaker pointed out that the parable of talents reminds us that everyone of us has a talent, or maybe only one-half a talent, but we each have something we can give.

Whatever God has given to us we must give back to others or it is lost to everyone. Look around you. There is someone hurting or lonely around you everyday. Why not light up his or her darkness with a smile, a friendly word, or a hug?

This little light of mine,
I'm gonna let it shine.

I Can Run!

Several years ago we worshipped in a small church in New England. The opening words in the church bulletin spoke to me in a special way that morning. They were from the pen of Isaiah:

They shall mount up with wings like eagles, they shall run and not be weary, they shall walk and not faint.

—40:31 (RSV)

This has always been one of my favorite verses in the Bible and one which I claim daily. Following my hip surgeries, when I was having difficulty accepting less than the best recovery, a friend sent me a note with the above Scripture verse and then added these words, "There are *many* ways that we run!"

I have kept the note in my Bible and those words have often helped me move beyond a discouraging time. In rereading this verse again and again, I realized that there was a truth here that would enrich my life, so I asked myself the underlying question: "What are the ways in which we run?"

For many years it has been impossible for me to run physically. But I have discovered that there *are* many other ways that I can run, and when I do, I can give you a gift also.

I don't have to run physically, but I can run by giving you a smile. We are told that it takes fewer muscles to smile than to frown. A smile can be my gift to you.

I don't have to run physically, but I can run by keeping in touch with you. Just a few lines in my handwriting tells you that I am thinking of you. A caring note can be my gift to you.

I don't have to run physically, but I can run by baking a loaf for you. That tangible bit of caring will help us both. A loaf with love can be my gift to you.

I don't have to run physically, but I can run by letting you
know that I am concerned when you are hurting. Just
"being there" and holding your hand can be my gift to
you.

Yes, there are many ways that we can run. And it is great that
they do not make us weary, but they do help us mount up with
wings as eagles. Only then, do we soar, even as the eagle, above the
mundane and the emotional and physical hurts that life inevitably
brings to each one of us.

Smiles: Your Best Gift

While sitting in the waiting room of a doctor's office, I
picked up a copy of the *New Yorker* magazine. As I idly
thumbed through its pages, an ad caught my eye. It was
for a designer's signature umbrella. Alongside a picture of the umbrella
were the words, "Let a smile be your umbrella." How unique!

It set my mind to thinking about smiles and umbrellas. In
general, we need umbrellas on days when the dark, heavy rain
clouds hang low and touch us with their drops. In general, we need
smiles on days when the dark, heavy clouds of life hang low and
touch us with their hurts.

So, how do we become smile-makers for *all* the days of our lives,
both dark and sunny? Proverbs 15:13 tells us, "A happy heart makes
the face cheerful . . ." (NIV).

This philosophy translates into smiles. And that is the secret.
We reflect in our faces that which we hold in our hearts.

On my kitchen counter I have a flip-over calendar of daily thoughts. The calendar is titled, "I'm Too Young To Be This Old." On May 11 the thought was: "A smile is a light in the window of your face to show your heart is at home." When we smile at a person we are saying to that him or her, "My heart is at home to you."

It does not take a lot of muscle to give the heart a lift. Someone has written these thoughts about smiles:

All people smile in the same language.
Your radiant smile can light up a room as you walk into it.
A smile is a curve that helps set things straight.
If someone seems too tired to give you a smile, just give him
    or her one of yours.

What other gift do you and I have to give that does not require shopping, does not need to be gift wrapped, and always fits?

Pass It On

There is no such thing as a self-made person! We are each a part of that which has gone before us—from our Judeo-Christian heritage well before the days of Christ down to what touched us yesterday. We are a part of it all and it is all a part of us. Even the "begat" passages in the Old Testament help trace the lineage of our spiritual ancestors.

We are *who* we are because of that part of the global world in which our ancestors made their home. Whether our grandparents were Asians, Africans, Europeans, or Native Americans, their customs and backgrounds have helped mold you and me.

I grew up in a generation where we did not talk about the "generation gap." We listened to, learned from, loved, and respected our great-grandparents, our grandparents, and our parents and their peers. They have all helped shape the persons we are today.

Sometime ago I attended an illustrated lecture by Tasha Tudor, a well-known writer and illustrator of children's books. What a very special lady she is! She spoke to us with refreshing spontaneity in her genteel Bostonian English.

She shared some of the many ways that she was influenced by her mother, a noted portrait artist. She told how her mother frequently washed her paintbrushes out in the bathtub as Tasha and her younger brothers took their baths at the end of the day.

Tasha acknowledged that her mother's talent and free-spirited lifestyle became a driving force in her life. Through the years she has given expression to this in the precious children that she has created with her own paintbrush.

Just as many tributaries flow into a river and determine its depth and width, so many persons have contributed to our lives and have helped shape us into the kind of persons we are.

Amanda Bradley has put it this way:

There are people who can't be forgotten,
Friends who are so much a part
Of all that you've done,
Of all that you've become . . .
They'll always be close to your heart.

Life is so devoid of meaning apart from relationships. The influence of others has played such an important role in our own development. We cannot divorce ourselves from the past if we really want to celebrate our lives in the present.

Just as we all receive so much from those relationships, so we, too, are contributing much to the lives of others around us. I wonder what would happen if each of us accepted the challenge of the apostle Paul to try to "outdo one another" in demonstrating acts of love and kindness to everyone whose lives we are touching today and will be touching tomorrow?

Have a Happy Brightday!

I have just unfolded a colorful placemat which I brought home from one of yesterday's trips. I was so impressed by the message on the mat that I have kept it and reread it many times:

> Those who bring sunshine into the lives of others
> Cannot keep it from themselves.

There are sunshine makers all around us. They are those who live life with a smile.

When she was a small child, my granddaughter, Amy, often made the birthday cards she gave me. On one of those cards she had drawn and colored a bright smiling sun. Over the face of the sun she had printed the words, "Have a Happy Brightday!" At first glance I thought she had made a mistake in the spelling. Then, I realized that was really her birthday wish for me.

I have thought of persons who have helped me have brightdays. I realize they have all had one thing in common—each person has shared a smile with me.

We tend to react or respond to many stimuli about us. We react to negative signals and respond to positive ones. How great it would be if everyone would always give off those upbeat signs of happiness!

Mother Teresa is one of the finest Christians of our day. 'Midst the smells, squalor, and suffering she is a prolific sunshine maker. She has encouraged others with these words:

> Smile at each other. Smile at your spouse. Smile at your children. Those smiles will help you grow up in greater love for each other.

Someone else has suggested, "Good friends are only a smile away."

Smiles beget smiles. It gives one a lift to smile at an office receptionist, a service station attendant, the clerk in a department store, or your child's teacher. Then, watch them as they respond, sometimes very hesitantly, but with a smile and a lighter step.

A friend sent me a bookmark printed on bright yellow paper, showing a "happy face" along with these words:

> Mark your book with a smile.
> It's easy to be pleasant when
> Life goes by with a song.
> But the person really worthwhile
> Is the one who can smile
> When everything seems to go wrong.

Have a Happy Brightday!

Donkey Givers

Not long ago, I heard a pastor challenge the members of the congregation to become dedicated "donkey givers." His message was based on Matthew's account of the events leading up to the triumphant entry of Jesus into the city of Jerusalem, and reads as follows:

> As they approached Jerusalem and came to Bethpage on the Mount of Olives, Jesus sent two disciples, saying to them, "Go to the village ahead of you, and at once you will find a donkey tied there, with her colt by her. Untie them and bring them to me. If anyone says anything to you, tell them that the Lord needs them, and he will send them right away."

> This took place to fulfill what was spoken through the prophet: "Say to the Daughter of Zion, 'See, your king comes to you, gentle and riding on a donkey, on a colt, the foal of a donkey.'"

> The disciples went and did as Jesus had instructed them. They brought the donkey and the colt, placed their cloaks on them, and Jesus sat on them. A very large crowd spread their cloaks on the road, while others cut branches from the trees and spread them on the road. The crowd that went ahead of him and those that followed shouted,

> "Hosanna to the Son of David!"

> "Blessed is he who comes in the name of the Lord!"

> "Hosanna in the highest!"
>
> —Matthew 21:1–9 (NIV)

The Scripture tells us nothing about the owner of the donkeys, but we do know that he responded immediately to the disciples' request to borrow them. Jesus mounted the colt and made His triumphant entrance into Jerusalem. This passage is really a story about donkey giving—giving what we have for His use.

There was a tired, hungry milling crowd on a hillside. A little boy gave his lunch—five hamburger buns and two large sardines. Jesus took them and fed five thousand men plus the women and children who were present. When all had eaten, there were twelve baskets left over. Five hamburger buns and two sardines are never enough until we give them away.

Often we fault the Bethlehem innkeeper for putting Joseph and Mary in the stable, which was probably a cave. But we must remember that he gave the only gift he had left that night—the cold cave with the smell of animals and straw. God took it and it became the birthplace for His Son, our Savior.

The challenge to each of us is to become a donkey giver. We can each give what we have to improve the lives of others.

Our gift may be a radiant smile. God can take it and use it to warm the heart of someone who is discouraged.

Our gift may be a warm welcoming home. God can take it and use it to restore someone who has been estranged.

Our gift may be the art of being a good listener. God can take it and use it to give confidence to one who feels inferior.

Our gift may be the art of teaching. God can take it and use it to inspire another to try again.

The season for gift giving is yearlong. Whatever we give to others in His name is never lost. He will take our gifts and multiply them far beyond our wildest imagination. Give your *best* gift.

Hands

Shortly after the close of World War II when we were in Europe, we were encouraged to visit a church in a small French village. The church had just been rebuilt. As we entered the sanctuary we were amazed by the grandeur of the interior: the exquisite stained glass windows, the gleaming white marble altar, and the tall graceful columns which stretched upward. Everything seemed to be in perfect taste and harmony.

Then, we came upon it! A strange statue of the Christ. All of the other images and art were strikingly beautiful. But this statue seemed to be so simple, almost crude. More amazing still, it had no hands.

We approached to within a few feet of the statue. We saw a bronze plaque at the base of the statue with these words inscribed on it in several languages:

You Are My Hands

The dynamic Christian from Japan, Toyohiko Kagawa, expressed a similar thought so beautifully in a poem he called *Discovery*:

I cannot invent new things
Like airships which sail on silver wings;
But today a wonderful thought
In the dawn was given.
And the stripes on my robe,
All shining from wear, were suddenly fair.
Bright with a light falling from heaven . . .
Gold and silver and bronze,
Lights from the windows of heaven.
And the thought was this:
That a secret plan is hid in my hand;
That my hand is big, big, because of this plan.
That God, who dwells in my hand

Knows this secret plan
Of the things He will do for the world using my hand.

What we do with our hands always reflects that which is in our hearts.

Hands are symbolic of service. Just now look at your hands and be thrilled by their movement and their usefulness. In order to receive the work that God would place in our hands, we must open our hearts to Him as well. He will not place anything in a folded and limp lethargical hand, in an uncaring hand, in a clenched, angry hand, or in a tightly gripped, selfish hand. He wants to use hands which are open in love and service to those in our families, in our churches and communities, and ultimately to those around the world.

Surely, members of our families are most sensitive to the many things we do with and for them everyday. When these acts are accompanied with a sensitive touch or a loving caress they take on added meaning.

Neighbors, friends, and many others about us are lonely and hurting. Many are nursing deep hurts and aching pain within their very souls. As we reach out to them with loving acts of kindness or show our caring concern for them in other ways, we become Christ's hands.

Throughout the world there are persons who are crying out for relief from their suffering, their hunger, and their fear. We cannot touch them physically. But in the name of Christ, we can share the fruits of our labor—our money and other resources—to fill their empty stomachs, bring healing to their bodies, and help calm their fears.

Two hands outstretched on a cross of wood,
From cruel nail wounds red with blood,
Bringing a lost world back to God.
O hands of my crucified Christ divine,
Take into Thine own these hands of mine
And teach them to serve with a love like thine.

—Mrs. Paul Moser

Just a Sparrow

Just a Sparrow

Several years ago, we moved into a new home in an area that was devoid of trees. We hung out our bird feeders in hopes that we could entice some birds from a nearby wooded area to come.

A few weeks after we had hung the feeders, our grandson, who was seven years old at the time, came for a weekend visit. One morning as I was getting breakfast, he called excitedly to me, "Granny, there is a bird out here at your feeder. Come quick and see it!"

I moved slowly and carefully to the window, hoping not to frighten the bird away. When I saw the bird, I said to my grandson, "But Hart, that is only a sparrow!"

The look he gave me revealed his deflated feeling about my reaction to the feathered friend he had found exciting. Then, his words hit home as he replied, "But Granny, God made the sparrow too. And He loves it just like He loves me and you!"

I have carried his candid words in my heart ever since that visit. I know I am God's child and He does love me and watch over me.

This truth has led me to seek out the words of an old gospel song that I learned as a child:

Why should I feel discouraged
Why should the shadows come
Why should my heart be lonely
And long for heaven and home?

When Jesus is my portion
My constant friend is He
His eye is on the sparrow
And I know he watches me.

Let not your heart be troubled
His tender words I hear
And resting on His goodness
I lose my doubts and fears.

Though by the path He leadeth
But one step I may see
His eye is on the sparrow
And I know he watches me.
          —Mrs. W. Stillman Martin

God made the sparrow. God made you and me. He does care
for each according to our needs. His eye is on the sparrow and I
know He watches me—and you too!

Where They Know My Name

O ur culture revolves around the numbers we have been given: a Social Security number; a license plate number; a driver's license number; a telephone number; a credit card number.

When we order from a gift catalog, the operator asks for a number on the catalog address label. When we call the newspaper about our delivery service, the operator asks for our telephone number. Our church offering envelope carries our church identification number.

In this day of computeritis, we all need to be known by more than the number on a card. We want to be called by our names.

I have been intrigued by the words of a song which opened a weekly sitcom:

Makin' your way in the world today
Takes everything you've got.
Sometimes you want to go
Where everybody knows your name
And they are always glad you came.
You want to be where everybody
Knows your name.

In our church, we sign an attendance registration form and pass it down to the end of the pew and back again. This enables us to know those sitting in the pew with us so we may greet them by name at the end of the service—visitor and member alike.

We use the same pharmacy over and over again where the pharmacist calls us by name. The dry cleaner gets our repeat business because the attendants know our name. We enjoy going to the same hair salon because everybody knows our name.

During a recent visit in one of our churches in a large metropolitan area, a minister told us of a man who came to church for the first time but left before anyone learned his name. When he came the next Sunday, the minister was able to greet him and learned his name was Jack. Jack came back to worship the third time and the pastor greeted him by calling his name. That afternoon, the minister's telephone rang and she recognized Jack's voice on the other end of the line. Jack told the minister that he was very depressed and had been thinking of committing suicide. But he said because she had called his name that morning at church, he sensed that she was a person who cared and he wanted to come and talk with her about his problems.

We all want to be in places where somebody calls our name and is glad we are there.

"You Are Something Special"

S everal years ago, our granddaughter accompanied us on a weekend trip. As we rode along for several hours together in the car, she taught me the words to a dear little song that she had learned in her church choir:

> I'm something special.
> I'm the only one of my kind.
> God gave me a body and
> A bright and healthy mind.
> He had a special purpose
> That He wanted me to find.
> So, He made me something special.
> I'm the only one of my kind.

In these days when you and I have become an anonymous number, it is refreshing to be reminded that we are "God designed, one-of-a-kind."

We are a part of the masses of this world, but we are very unlike, unique individuals. A friend of mine speaks of this as "God's amazing diversity."

Sometime ago in his *Better Homes and Gardens* column, "The Man Next Door," Burton Hillis wrote of the beauty of these differences. He reminded us that we can never tell the beauty of a snowflake by looking at a snowdrift. It is only as we examine the individual snowflake under the close scrutiny of a magnifying glass that we see its unique beauty. That beauty is not found in any other snowflake.

One line of the little song suggests that we each have a special purpose that God wants us to discover for ourselves. God does see the possibility in each one of us. He wants us to bloom where we are planted.

Someone has said, "Jesus never met an unimportant person." Think of the "ordinary people" His life touched intimately. Because His life touched them, they became very special.

You are something special! And that is the secret, my friend.

Thorns among the Roses

Thorns among the Roses

It is a bright and beautiful summer morning. It is the beginning of the season when I sit on our covered deck in the cool of the morning before the heat of the day sweeps across and envelopes it.

My husband is a rosarian. I am looking out over our rose garden which just now has over one hundred beautiful plants in full bloom. The blossoms range in color from the pale pinks of a delicate Haviland plate, to the soft peach of an art glass peach blow bowl, to the deep apricot of a bride's overlay basket, to the velvety reds of a cranberry glass hanging lamp.

And to think God created all these gorgeous colors long before the early artisans made Haviland china, peach blow art glass, colored overlays, or elegant cranberry glass pieces.

Through the years the rose has been an inspiring metaphor of life. The sharp thorns are there, yes, but what fragrant beauty and aesthetic joy blossoms forth just beyond the thorns. The thorns on our roses are real. When I arrange the flowers, I use a dethorner which a friend gave us. However, even then I often cut myself on some of the sharpest and thickest thorns. The cuts hurt and draw blood.

The words of the country ballad often come into my thoughts, "I never promised you a rose garden." One of our roses is named Lynn Anderson in honor of the singer who made that song popular.

Life brings a number of "thorny experiences" to each of us. There are times when we are discouraged and defeated in spirit and we feel the cuts and bruises.

We all know that life is not a rose garden with blossoms on stems without thorns. Could it be that the hurt from the thorns makes us more sensitive to the needs of others? Perhaps it is only as we, too, hurt that we become more capable of empathizing with Christlike love for those who are hurting.

Our loving God wants only good things for all of us. But because He has given us and others the freedom to make many choices during our lifetimes, there are some thorns among the roses.

As I continue to look at those long-stemmed beauties, I am reminded of the beauty of God's faithful promises to us:

> Underneath are His everlasting arms that cradle us; deep within us is the Holy Spirit who guides and comforts us; walking alongside us is the Christ who fills us with the joy of His presence which He promised when He said He would be with us to the end of the world.

I am so grateful that we may all claim those promises, especially when life seems to be filled with sharp thorns on our roses.

Catalogs of Hope

*T*iny pellets of sleet are pounding on my kitchen window. I hear the snap and crackle of the hot fire in the fireplace as I sit with a pile of assorted seed catalogs in my lap. During the last few weeks, the postman has been leaving them in our mailbox every few days.

From the brilliant yellow marigolds in the *Burpee Garden* catalog, to the delicate lavender clematis on the *Wayside Gardens* cover, to the deep coral-pink Jackson Perkins rose of the year, they all have one thing in common: each catalog promises pages of beautiful colorful blossoms in the months to come. Through the years they have become my February *hope books:*

Hope—What Is It?

Hope involves our ability to look beyond the present to the future.

Hope is the emotion which enables us to project ourselves from a bad situation to a better situation.

Hope is a feeling that urges us to "keep on" even in the midst of adversity.

Hope is the faint light at the end of the dark tunnel.

Hope invites us to dare to claim a promise.

Hope—What Does It Do?

Hope will help us look for the rainbow even when it is raining.

Hope will help us fight a life-threatening disease in our bodies.

Hope will give our mind, hands, and feet a job to do which we thought we could not do.

Hope will keep a longed-for dream alive in our hearts.

Hope will make the promises of God possible in our lives.

The sleet is still hitting my window, the fire is still crackling at my side, but these seed catalogs in my lap have lifted me beyond the storm.

Hope springs eternal in the human breast.

Notes and Rests

*T*his bright morning the pungent smells of a spring day creep in the open door and envelop me. Only the soft rustle of the wind and the scolding cardinal at the empty bird feeder break in on my forced solitude. I am alone.

I know that not far away are the busy thoroughfares of our city, and the many ribbons of intertwining interstate highways. All of these are alive and vibrant with the motion of persons, persons, rushing, rushing.

But sitting here alone, I feel apart and a stranger to the mainstream of living so close by.

The sound of the birds' songs outside reminds me of the musical orientation of my life—the mass of notes and rests through long hours of practice sessions that were once a part of me. Without the rests, the mass of notes lose their beauty. I am also thinking of the notes and rests that life brings to each of us. The notes form the melodies and harmonies, the moving part of our existence, while the rests bring the pauses, the waiting times, when the tempo of our lives moves from allegro to andante, and finally comes to the rest that life brings to each of us.

It is always difficult to make the rests in my life fulfilling and creative. A friend once wrote, "God orders the stops as well as the steps of a good man or woman."

We have been accused by those in other world cultures of being a nation of activists, a people driven to be in constant motion to find meaning. What then, of the times in our lives when circumstances slow us down and seem to cut us off from the fresh, free-flowing stream?

Many phrases flood into my thoughts this morning:

Be still and know.
He leadeth me beside the still waters.
Wait patiently for the Lord.

Dear God: Help me to break down the walls of my aloneness so that I may know oneness with you.

Please assure me that these days of rest are so ordered by you just as you have set the notes of my life in motion through the years.

Help me to know again the vibrancy and purpose of life. Help me to find new ways to use these quiet hours creatively as I grow closer to you and become more sensitive to those about me. Amen.

Take Time to Smell the Roses

This is my Father's world,
And to my listening ears
All nature sings and round me rings
The music of the spheres.

This is my Father's world,
I rest me in the thought
Of rocks and trees, of skies and seas
His hand the wonders wrought.

This is my Father's world,
The birds their carols raise,
The morning light, the lily white
Declare their Maker's praise.

This is my Father's world,
He shines in all that's fair,
In the rustling grass I hear Him pass
He speaks to me everywhere.
                    —Maltbie D. Babcock (1854–1901)

We live in a wonderful world . . . a world in which God has given us not only the necessities of life, but so many things of breathtaking beauty. God speaks to us through the sights and sounds of loveliness He has placed in this world for us to enjoy.

Two persons may walk down the same street, may stroll the same path through the woods, or visit the same quiet garden, and receive two completely different messages. To the one, engrossed in his or her own thoughts and problems, perhaps there will be no message at all. To the other, in tune with God, there will come a thrilling renewal of faith:

- from the song of a bird;
- from the rustle of a leaf caressed by the wind;
- from the strong smell of damp earth; or
- from the sweet fragrance of a rose.

This is rose-blooming time in Tennessee, and I have made a new promise to myself—I will take time to smell the roses!

Slow me down, Lord! Ease the pounding of my heart by the quieting of my mind. Steady my hurried pace with the vision of the eternal reach of time. Give me amidst the confusion of my day, the calmness of the everlasting hills. Break the tensions of my nerves and muscles with the soothing music of singing streams that live in my memory. Teach me the art of taking minute vacations—of slowing down to look at a flower, to chat with a friend, to pat the neighborhood dog, to read a few lines of a good book. Let me look upward into the branches of a towering oak and know that it grew because it grew slowly and well. Slow me down, Lord, and inspire me to send my roots deep into the soul of life's enduring values.

—Author Unknown

Someone has said, "Only that day really dawns to which we are alive." Recently, a dear friend who is ill wrote that she takes nothing for granted. She enjoys all the little sights and sounds of each day.

I must take time to live, not just exist. I will not hurry through this day. I *will* take time to smell the roses. Are there also some roses where you live?

A Bit of Sky

Someone has said, "The bluebird carries a bit of the sky on his back." Surely, it is the brilliant blue of the sky that the male bluebird displays as he preens in the bright sun.

Bluebirds have found us again. We first saw the pair in late February. They sat on a high pole bird feeder scouting out the bluebird box.

The bluebirds live in small family flocks during the winter months. Then, in January and February, the larger family scatters in search of their own homes to mate, to build a new nest, and to settle down and raise several new families. They often raise as many as three or four broods during a summer season. Those from the first hatching usually stay with the parent birds to help feed their siblings in the second and third broods. And so, the cycle continues on into the next season.

We watched them daily as they used some of the pine straw from our rose beds nearby. After the nest was formed, they seemed to hunt for softer material to line the inside of it.

A few days later, we checked the box and there were six tiny blue eggs in the nest. Then, came the time for the female to sit on the eggs. The male regularly came to feed her.

Sometime later, we checked the box and there were six tiny birds. These days we have enjoyed watching both parents feed their babies and fend off other birds—some even as large as robins and mockingbirds. We are hoping that we can be aware of the day when the mother will coax them out of the box and will fly alongside them until they land safely in the grass.

They are truly a wonder of God's creation. They seem to have a built-in ability to survive. In recent years, they have been able to adapt from life in the open country to life in the woods or even nests in suburbia.

We call these pretty friends our "bluebirds of happiness." They remind us of God's faithfulness and order throughout the universe He has created. They also foster feelings of happiness and contentment with the simple things of everyday life. Somehow they call to my mind:

- the earthy smell around us after a refreshing rain;
- the tender caress of a loved one;
- the smell of freshly baked bread;
- the glow of lights in our homes at dusk;
- the moving harmonies of a symphony;
- the smell of freshly cut grass;
- the hugs of grandchildren when they come to visit;
- the beauty of the first blossoms of spring;
- the letter in the mailbox from a dear friend.

Oh, "blue bits of happiness," come again and again.

Queen Anne's Lace and Other Weeds

*J*uly is one of my favorite months. It is the season that God has chosen to adorn the roadside and fields with Queen Anne's lace and other weeds.

At least once every summer, we drive out along the country roads and gather Queen Anne's lace, elderberry blossoms, wild daisies, and the lovely fragrant milkweed blossoms.

As I arrange the armload of beauty in a favorite antique container, I feel I am looking at a reflection of my Creator:

> Only God could make such beauty bloom with so much abandonment;
> Only God could enable anything to flourish in such arid, sometimes rocky, and unproductive places;
> Only God could create such regal long-stemmed beauties with minute lacy intricacies;
> Only God could make roadside weeds so beautiful.

God rarely reveals Himself to us in the blazing spectacular things of life. He came to us as a baby in a manger. He has taught us through leaven in a lump, a widow's mite, and a mustard seed. He continues to come to us in the simple beauties of this world, in the everyday routines, and through our relationships with others.

Knowing of my fondness for Queen Anne's lace, a dear friend sent me these favorite lines:

> He called it just a common weed
> Wild carrot with its wayward seed
> She put it in a lovely vase
> And gently named it "Queen Anne's lace."

We do not need wealth to buy a ticket to enjoy God's beauties along the roadside. We do not need prestige and fame to be admitted into God's kingdom. We need only to open ourselves to Him and to His will for us. He can take our weedy, wayward lives and transform them into bits of loveliness which are just as beautiful as Queen Anne's lace.

Roses and Cabbages

F once read, "Every home needs a bouquet as surely as it needs a cabbage." (I would wish for roses in my bouquet.) In other words, we each have a deep need for the beautiful as well as the need for bread.

In today's pragmatic world where only what is practical or profitable seems to be worth keeping, some would say that we only need cabbages, for they are essential for feeding our bodies. So, our educational systems are attaching increasing importance to the math and science tracks while giving less attention to the importance of art and music. This developing trend is a cause for real concern.

What then of the things in our world that feed our souls? Are they to be cast aside?

We have reminders that God created unlimited beauty in this, our world. We think of all of the natural beauty He has given to us: a welcoming sunrise in the morning; a brilliant bird at the feeder; the wild flowers along a stream; a snowcapped mountain; and a placid New England lake reflecting the brilliant foliage in autumn.

We also think of the beautiful things around us that women and men have created: an attractive home; a majestic Gothic cathedral; the warm patina of a newly refinished antique cherry table; the perfectly proportioned modern skyscraper; a beautifully designed and manicured formal garden; and the fields of green rustling corn and golden waving grain.

My husband grows roses. Because he is an early morning person, he works in his rose garden before I am up in the morning. Many mornings from mid-May to late October I will find beautiful roses on my kitchen work counter when I come out to get breakfast. He is "seasoning" them to be arranged later in a favorite antique cut glass bowl, or to be placed in the refrigerator to be taken to a shut-in, a friend, or a neighbor. On such mornings my soul is surely fed by the beauty of those roses.

Our lives are greatly enriched by the many art forms around us. They deepen our sensitivity to the beauty in God's world. Who of us has not been:

- touched by a beautiful Renoir painting?
- challenged by thought-provoking verses of Robert Frost which we memorized when we were in school?
- stirred by the familiar strains of a Beethoven Symphony? or
- moved to tears by the melody of a hymn which we associate with an important point in our spiritual journey?

What a fantastic treasure God has given us when He gave us the gift of sound and the ability to hear. Music, perhaps more than any other art form, plays on the chords of our innermost emotions to move us, even comfort us, in some of our deepest and most profound life-changing experiences.

Music in all of its varied forms:

- can challenge us to courageous acts of patriotism;
- can motivate us to reach out to help those about us who are hurting; and
- can inspire us to reach higher in our quest for a closer walk with our God.

It is important for us to recognize and experience the beautiful in our lives. The ability to experience and appreciate beauty sets us apart from all living things. Perhaps that is what it means to be a living soul.

We do well to remember that we do not live by bread alone. We live by beauty as well as by bread. Bread sustains our existence. The beautiful transforms that existence into abundant living.

Bouquets are God's "bread from heaven." We *are* fed by roses as well as by cabbages.

A Summer Song

I have just heard a beautiful concert. I was not sitting in a padded seat in our symphony hall, but in a wicker chair on our covered deck on a lovely summer day. This was God's concert for anyone who would take the time to listen.

My neck is strained, not from looking up from the front row to see the orchestra, but from gazing up to the top of the chimney on our neighbor's house. I watched as I listened, to a mockingbird as he sang his heart out.

My words cannot possibly describe the beauty of his songs. Sometimes, the sounds were like the throaty silken melody of a violin. At other times, I was reminded of the plaintive haunting call of the oboe. And again, he would break forth in a shrill crescendo of the intricate trills and glissandos of a flute or a piccolo.

Ornithologists say that the mockingbird can expertly imitate over forty species of birds, in addition to making numerous call notes and other sounds! They will often sing all through the night, especially when the moon is shining.

Even though the bird's song sounded happy, he flitted restlessly out from his perch and back again. He seemed to be searching for a better vantage point or a place that offered him greater security.

How much we are like that mockingbird. We put on a happy face, but we, too, sometimes flit needlessly from job to job, from fad to fad, from craze to craze, in search of a better vantage point or a greater sense of security.

We forget that God is there all the time and He is waiting for us to turn to Him. We tend to forget that He will help us find purpose and peace in our lives. So it was that the sainted Augustine concluded that "Our hearts are restless until we find our rest in Thee."

A Beautiful Rambling Rose

*J*t was a beautiful, crisp morning in northern California. The giant redwoods lined our road on either side, making lacey moving designs on the road ahead as the sunshine filtered through the branches.

As we rode along, my thoughts were centered on those majestic trees which reached hundreds of feet heavenward. As some of them were two thousand years old, I realized that they had stood in this place at the same time that Jesus Christ had walked in another place on this earth. How awesome!

When we passed through the small village of Jenner, near the sea, we came around a curve in the road and my eyes were drawn to "it." 'Twas the most beautiful rambling rose over the doorway of a paint-peeled, abandoned, dilapidated old building with a sagging roof. What a picture for some artist to paint!

I quickly reached for my notepad in the seat beside me to try to paint with words some of the thoughts which flooded over me when I saw the rambling rose. That old-fashioned rose was unashamedly "blooming its heart out" on that old weather-beaten building. Those surroundings were surely not conducive to its giving forth such beauty. But the rose did not know the building was worthless. An old saying came to mind: "It just bloomed where it had been planted."

The strength and beauty of that rose came not from that old building, but from deep down in God's good earth. The nutrients from the soil entered the root system and moved up through the twisted canes of the rose plant in order to form those beautiful buds and blossoms. The spectacular beauty of that rambling rose was generated from its source—God's earth.

Life brings to each of us a different set of circumstances:

Some have been born into beautiful homes filled with love—
others into dysfunctional homes marred by constant strife.

Some have been born into material wealth—others into abject poverty.

Some have been born with strong healthy bodies—others with lifelong handicapping conditions.

Some have been born with genes that produce brilliant minds—others with genes that produce minds that are not as sharp.

But to everyone and everything God has created, He has also given the "power to become." This is our birthright! We have the innate power to become growing, caring, serving contributors to a more beautiful world.

Let us "bloom our hearts out for the Lord" wherever He has planted us. Can you think of a greater challenge? I cannot.

Golden Days

September and October bring "golden days" in New England. There is absolutely no sight to compare with the bright crimson red of the hard maple trees that grow in abundance there. Along with their vibrant colors, add the verdant green of the stately pines and the thin shafts of the white birch trees all on the same hillside, and you have God's incomparable painting of Vermont.

Next, pull down the bright sunshine to those hillsides with their green pastures below and you have the makings of a golden day. Each curve in the narrow country roads opens up a panoramic view more breathtaking than the one you have just passed.

While basking in one of those golden days, it occurred to me that there are other kinds of golden days we experience. A golden day produces warm feelings which are engendered in our hearts by something beautiful—something we see or something we feel . . .

A golden day can envelope us when we learn of the Christian service of a grown daughter or son.

We know a golden day after a special conversation with deep meanings with a teenaged granddaughter.

Golden days warm us after spending some quality time with a friend who is ill.

A serendipity golden day rushes in with an unexpected visit from a childhood friend whom we have not seen for many years.

A golden day soothes us during the luxury of some hours spent with a good book.

A golden day always comes with a letter from a friend which links us once again with his or her life.

A golden day is born for us during our prayer time for those on our daily prayer list.

A golden day ends with the lowering sunset with rose and pink shades we enjoy from our deck as the sun drops down behind our Tennessee hills.

Yes, not all golden days are in Vermont. Thanks be to God! He grants them to us wherever we are. All we need is a sensitive heart to recognize them.

When Life Gives You Scraps

When Life Gives You Scraps

Almost every week a number of interesting catalogs arrive in our mailbox. They show pretty clothes for the new season and interesting gifts to order for my family and my friends.

One of the catalogs I receive always has very attractive items for the home. The sayings on two little plaques in the last catalog caught my attention. One read:

If life gives you scraps
Make a pretty quilt.

The second one read:

If life gives you a lemon
Make refreshing lemonade.

Barbara Johnson's life has not been easy. A tragic accident left her husband blind and crippled. She lost two sons—one in Vietnam, and the other on a highway in the Yukon when a drunken driver plowed into his car. Her third son completely disappeared for nine years when he pursued a homosexual lifestyle.

Yet, she was able to write an upbeat and positive book which she titled, *So, Stick a Geranium in Your Hat and Be Happy!* (Word Publishing Co., 1990.) She wrote:

101

Life isn't always what you want, but it's what you've got, so can you stick a geranium in your hat and be happy? I know you can, no matter what happens. We all have to endure troubles in life. Sometimes we may go along for awhile with just common irritations and then, WHAM! A big problem hits, and we are thrown into a real valley experience. But I believe you grow in the valleys for that's where the rich fertilized soil is.

Somewhere, someone called my attention to an insightful observation by St. John of the Cross, who wrote:

I am not made or unmade by things that happen to me, but by my reaction to them. This is all God cares about.

Being a Christian or being a good moral person does not insulate us from the trouble spots of life. Rabbi Kushner struck a responsive chord for many of us when we read his book, *When Bad Things Happen to Good People* (Schocken Books, 1981). He illustrates over and over how life is not fair.

Life was not fair to the composer, Ludwig von Beethoven, when he lost his hearing. But he wrote his greatest composition, *The Fifth Symphony*, after he was totally deaf.

Life was not fair to a child, Helen Keller, who was born without two crucial senses—hearing and sight. But she left a trail of inspiration for persons throughout the world.

Life was not fair to Jane Merchant, a young woman in East Tennessee, for whom even a slight movement often caused her brittle bones to break. But the sensitive poetry she penned while lying in bed won her the title of Poet Laureate of Tennessee.

Phillips Brooks, the great hymnist and preacher of the last century, wrote: "The truest help which one can render to a man who has any of the inevitable burdens of life to carry is not to take his burdens off, but to call out his best strength that he might be able to bear them."

We are reminded again of Barbara Johnson, who wrote of this concern out of her own struggles:

We can choose to gather to our hearts the thorns of disappointment, failure, loneliness, and dismay because of our present situation, or we can gather the flowers of God's grace, unbounding love, abiding presence and unmatched joy!

Christ never promised His disciples of His day, nor those who would follow Him in years to come, that we would have a painless, trouble-free existence. But He did promise to be with us and strengthen us all our lives, when He said, "And, lo, I am with you alway, *even* unto the end of the world" (Matt. 28:20 KJV).

Because of that promise and His presence, we can live triumphantly everyday!

Impossible Dreams

In one of his books, Frederick Beuchner says that we all need some "wild dreams" to stay alive emotionally and physically. He believes that things of importance seldom happen in our world without some wild unrealistic dreams in the minds of the men and women who eventually bring them about.

Many times we equate dreams with youth. But research has proven that middle-aged persons, and even the very elderly, will wither and finally dry up without the soul nourishment that comes from dreaming.

God marched across the pages of the Bible through the wild dreams of His people. For years Sarah and Abraham dreamed of having their own child. Despite the limitations of their age, God gave them Isaac—surely a wild dream come true.

Moses had a dream that someday he might lead the children of Israel out of their bondage, despite their long years of captivity and his own speech impediment. His dream came true when they crossed over the Red Sea on their way to the Promised Land.

Jonas Salk dreamed of a world free from the crippling disease of polio and God used Salk's intellect and research to help him realize his wild dream.

In his well-known speech at the foot of the Washington Monument, Martin Luther King Jr. told of his dream, and despite difficult odds, his words fueled the civil rights movement of the sixties.

Dr. William Charnley, a noted English surgeon, dreamed that someday there would be a prosthesis to relieve the crippling effects and excruciating pain of arthritic hips. He pursued his dream and today thousands of persons are free of their wheelchairs.

How rich our lives are because men and women have dreamed wild dreams!

Don Quixote, in the musical, *Man of La Manchia*, sings:

To dream the impossible dream,
To fight the unbeatable foe,
To give of my last ounce of courage,
To go where the brave dare not go.

This is my quest, to follow that star
No matter how hopeless, no matter how far!

God continues to change the world through persons like you and me who dare to dream wild dreams and then pursue them.

Have you chased after any of your wild dreams lately?

Change

Change is certain. Change is inevitable. Change is difficult. Many times we tend to view change negatively, for change shakes the very foundations of our security systems.

Several years ago, a television series, *Aaron's Way*, told the story of an Amish family. Due to unexpected circumstances, they found themselves transplanted from their simple, safe, and secure cocoon-like community in Pennsylvania, to the grape vineyard country of southern California. The show portrayed the extremely difficult changes that the move brought in the lives of this family of five. From buying their food in a supermarket, to riding on an escalator, everyday was full of unnerving changes and adjustments for them.

On one occasion, the husband asked his wife what her greatest challenge of that day had been. Her reply was, "Learning how to turn the timer off on the electric oven in the wall." Then, she added, "A wood stove is much easier to use."

Change in whatever degree or situation is threatening to most of us. There are many changes that come to each of us. Changes in:

- our health or in the health of one we love;
- our life work due to a change of position or the loss of a job;
- the economy which threatens our lifestyle;
- the morality standards in our culture;
- our family due to death or divorce;
- our relationship to a parent as our roles are reversed and we become the parent and our parent becomes the child; or
- our address which removes us from a familiar neighborhood and a supporting circle of friends.

A few years ago the latter was a major change for me. As winter gave way to warmer spring days, I missed our wooded hillside with its daffodils, forsythia, redbuds, and dogwoods. I wondered if the bluebirds came back again to build their nests in the boxes we left. I missed many of my friends in our neighborhood and church.

But I knew I must begin again. I was sure that blooming things would soon begin to appear in our new yard, and I knew I would soon begin to feel at home in our new neighborhood and church.

Perhaps God uses change to help us grow and to keep us from becoming stale, stagnant, and too "set in our ways." Perhaps it takes change to move us beyond what we are to what God knows we can become. Perhaps change transplants us from a comfortable and familiar hand-holding pattern of life in order to cause us to reach out to new hands, new relationships, and new situations.

Through it all we can take comfort in the thought expressed in the words of that old hymn:

> Change and decay in all around I see,
> O Thou who changest not, abide with me.

Simple Gifts

*T*he Shakers were a gentle people who came to America seeking religious freedom just before the Revolutionary War. They established eighteen Shaker communities up and down the East Coast and in Kentucky and Ohio. In 1855 the Society reached its peak population with 6000 members.

Their aim was to establish a celibate, utopian Christian community. Simplicity, order, and harmony were the tenets of their faith. For them, labor was a sacred commitment. Believers first and excellent craftsmen second, the Shakers were motivated by founder Mother Ann Lee's words, "Put your hands to work and your hearts to God."

Originally called the United Society of Believers in Christ's Second Appearing, they became known as Shakers because their frenzied singing and dancing were at the heart of their worship experience. This beautiful Shaker hymn, "Simple Gifts," reveals some things about them and their beliefs.

'Tis the gift to be simple,
   'Tis the gift to be free,
'Tis the gift to come down
      Where we ought to be.

And when we find ourselves
   In the place just right
'Twill be in the valley
      Of love and delight.

When true simplicity is gained
   To bow and to bend
We shan't be ashamed.

To turn, turn will be our delight,
   'Till by turning, turning
We come 'round right.

The Shakers believed that in turning, turning as they sang and shook, they were more open to God's will. They trusted Him to help them find just the right place. That was a mark of their obedience.

What of my obedience to God? How can I know with assurance that I am being obedient to God's will for my life? I want *to be* where God wants me to be, and *to do* what He wants me to be doing.

Many times it takes a lot of turning, turning for us to come 'round right. It means opening our hearts and our heads so fully to

God that we place our lives in His hands. It means listening and waiting for His voice.

He will lead us and sometimes shove us in the right direction. Surely, one of His greatest gifts to us is the assurance that we *are* where He wants us to be. Only in that way can our lives be rich and full of delight as we serve Him in the place just right.

Metamorphosis

Webster's classic definition of the word *metamorphosis* is "to change in form, to change in structure, to change in function." The word *transformation* is also suggested as a synonym.

Every living thing comes into this world with the capacity to become more than it is. The potential to change into something better is always present:

A drab lifeless seed can become a blossom.
An ugly caterpillar can become a beautiful butterfly.
A faltering colt can become a racing thoroughbred.
A helpless babe can become a capable adult.

All of these imply change—gradual, but sure change. As human beings we often drag our heels at any thought of change. We are so comfortable in our soft, safe, and comfortable little cocoons that we want to stay just as we are.

Change involves risk. Change involves vulnerability. Change involves the unknown. Because the results of change are not always known, we often tend to pull back from the newness and fullness of life that God would have for us.

God has given the "power to become" to every living cell. To us He has given even more. John puts it this way: "But to all who received him, who believed in his name, he gave power to become children of God" (John 1:12 RSV). He has given us that unique power to claim our divine inheritance.

What Would I Be?

God has His way with a butterfly
Fragile dweller 'twixt earth and sky
Lovely harbinger of the spring.
He tenderly fashions each gossamer wing,
Then stencils them with His own design
Intricate line upon fine, thin line.
Then, dipping His brush in the rainbow's hue
He colors them golden, and green, and blue
And sets them adrift in the warm spring air
Bits of loveliness floating there
Between the earth and the bending sky.
God has His way with a butterfly,
What would I be
If He had His way with me?

—Martha Snell Nicholson

Anxiety or Serenity

*T*wo words that we hear or use many times are anxiety and stress. A little trade paper that is delivered to our mailbox each week carries the column entitled, "Emotional Wellness." The subheadings have included such emphases as: Improving your self-image; Relaxation—controlling our anxieties; Stress—sources, symptoms, solutions; and Overcoming depression by breaking the cycle.

Webster gives us these definitions:

anxiety—worry or uneasiness about what may happen;
stress—pressure, especially outward forces that strain or deform.

These definitions suggest that anxiety is usually something that we engender from within ourselves and stress usually comes from persons or events outside ourselves. If this is true, it implies that we can do much to control anxiety and we need to learn how to cope with those things which produce stress.

Psychologists have concluded that 90 percent of the things we worry about never actually happen. They also remind us that anxiety and worry rob us of much of our creative energy. They can even immobilize us.

A thought for the day on the little flip-over book by my bedside reads, "Blessed is the one who is too busy to worry in the daytime and too sleepy to worry at night."

An old Chinese proverb asserts, "That the birds of worry and care fly about your head, that you cannot change. But that they build nests in your hair, that you can."

A close friend has an ever-recurring battle with anxiety. She is a perfectionist and strives to have everything "just right." She is inclined to dwell on an enormous total task and tends to be overwhelmed by it. She is a person who enjoys planning for events far down the road and strives to have all plans in place far in advance. You and I know that life usually does not afford this privilege.

In his book, *The Road Unseen*, Peter Jennings wrote:

> During the walk across America, I learned there was no other way to make it than to put one foot in front of the other and take the joys and the pains and the worries as they came. That way everything was much easier to handle.

> When we were crossing the mountain ranges of Colorado, I realized that I couldn't worry about who or what we would run into a week farther up the trail, because if I didn't concentrate on where we were at the time, there might not be that next week. If I had looked at only getting to the top of the mountain, I might have been overwhelmed and given up, but instead I took the mountain one step at a time.

During one of America's darkest hours, Abraham Lincoln raised the spirits of his aides with this story from his youth:

> When I was a small boy, I had a terrible fear of the dark. I always tried to get to sleep before nightfall. One night my father taught me this simple lesson: We were fixing harnesses in the barn, and my father asked me to go to the shed for more supplies. I stood at the barn door frozen in fear of the dark night. My father came up to me and said, "Pick up the lantern. What do you see?" "The oak tree," I answered. "Is there anything between you and the oak tree?" "No." "Then walk to the tree and lift the lantern again. . . . What do you see now?" "The mulberry bush." "Walk to it and lift the lantern again." By the time I'd gotten to the bush, I had figured out the procedure. So I made my way, step by step, from tree to bush to coop to shed and finally

to the supplies. It was a simple lesson, but it can take you a mighty long way.

We may ask God to lead us, but we must also have the courage to step out into the unknown without fear and with confidence in Him.

In Philippians 4:6–7 (LB), the apostle Paul writes:

Don't worry about anything; instead, pray about everything; tell God your needs and don't forget to thank him for his answers. If you do this you will experience God's peace, which is far more wonderful than the human mind can understand. His peace will keep your thoughts and your hearts quiet and at rest as you trust in Christ Jesus.

Passages

We are told that from birth to death life is a series of passages. We have learned from the experiences over the years that change and sometimes even trauma accompany our rites of passage.

From infancy to early childhood, through puberty and the adolescent years, there are points of passage which each bring about major changes. Then follows young adulthood and the mid-life crises we hear so much about. Finally, we pass into retirement and the years of older adulthood.

We have recently experienced retirement. Our routine lifestyle has suddenly been altered. With these changes has come the knowledge that we have left the work-a-day world of career and corporate involvement behind us. We are going through another major passage in our lives.

Henry Durbanville has written a charming essay about growing old gracefully. He suggests that God might ask him a hypothetical question, "Henry, would you like to turn back the calendar and relive any portion of your life from early childhood to your senior years?" Henry says his answer would be a definite, "No!"

Henry feels he has really reached the golden years. Because of his keener awareness of life:

- the flowers are brighter;
- the breezes are more gentle;
- the sunsets are more beautiful;
- the relationships are sweeter.

If we really believe in God's plan for us, every age is the perfect age to be. Years ago Jesus admonished us: "'Therefore I tell you do not be anxious about your life, what you shall eat or what you shall drink, nor about your body, what you shall put on . . .'" (Matt. 6:25 RSV).

Mark Twain once said, "Age is really a matter of the mind; if you don't mind it, it doesn't matter."

Wintertime

There are days when my soul
Knows the cold gray winter
Of doubt.

When each new day
Means just another
Painful beginning,

When joy and purpose
And usefulness seem
Beyond my reach,
When my heart cries
Out in anguish
"Where are you, O God?"

I wait—
I listen for your answer,
But the stark leafless
Trees stand mute,
Matched only by the
Silence of the damp
Lifeless sod.

I need you!
"Where are you, O God?"

Hope

An interesting "good news" item on television told of a new project in the subway trains of New York City. The panels above the windows on some of the trains are being used to convey a new and different kind of message. Some of the garish ads which promote remedies for everything from headache, to heartburn, to halitosis are being replaced. The new messages are an attempt to offer positive words of encouragement and hope to passengers to ponder as they ride the trains across that massive city.

New York, with so much violence—muggings, rapes, killings, and other forms of violent crime—is in desperate need of some positive influences, according to the reporter. He reported that one of the first efforts was to quote a poem from the pen of Emily Dickinson:

Hope is the thing with feathers
That perches in the soul,
And sings the tune without the words,
And never stops at all.

Feathers are light, wispy bits of elusiveness. Many times they are seemingly just beyond our grasp when we try to capture them in our hands.

In his book, *Love, Medicine, and Miracles*, Dr. Bernie Segal, a noted surgeon who has devoted his hands to working with cancer and AIDS patients, writes of hope. He tells of one patient after another who prolonged the length of their lives, enriched the quality of their lives, and even beat the odds of a negative prognosis by their sheer will to live. Hope, Dr. Segal believes, is rooted in the will, and fuels their determination to live.

Another story of hope was shown on a recent television newscast. David Sanborn, a young man from Tennessee, has been suffering from ALD, a debilitating disease, for the last several years. In an interview he said, "I am literally trapped in my body." Doctors had given him only a year to live, but he has survived for six years. His concluding words in the interview were, "You must keep a sense of humor about yourself and have the will to stay alive."

If, then, our will, that innermost force in our being, is the all important source of hope, then we must train our will to be a hope producer. We do have outside hope, helpers in our lives: the members of our family, our friends, our pastors, our doctors, and others who encourage and inspire us. However, in the final analysis, we must become the primary activators of our hope as we reach down deep into ourselves and rise up, as the Phoenix, from the ashes of our lives.

For the Christian, this means looking to God, the very ground of our being (will) and the ultimate source of our hope. It is as we bond ourselves with Him through our daily spiritual disciplines that our wills are strengthened and we *can* hold the "feather of hope" tightly in our grasp.

Who Am I?

Am I the one who seems to be happy and outgoing?

Am I the one who appears to be handling physical problems?

Am I the one who seems to be in charge of my life?

Am I the one who is composed in the presence of tragedy?

Am I the one who appears to possess power for living?

OR

Am I the one who knows plaguing doubt and painful struggle?

Am I the one who knows depressing hours of loneliness?

Am I the one who feels unsure of myself in new situations?

Am I the one who feels threatened in the presence of talent?

Am I the one who pulls back from deep relationships?

*W*ho am I? One or the other? Or, am I all of these and more? We all wear masks at different times in our lives. The many masks that we wear are like many facades on buildings. Just as a facade on a building does not tell us about the shop inside, so, the masks that you and I wear prevent those around us from knowing who we really are. We continue to wear masks of:

- gaiety when we are lonely;
- aloofness when we are insecure;
- anger when we are hurting physically;
- haughtiness when we are guilty;
- flippancy when we are embarrassed; or
- shyness when we are emotionally wounded.

The world is not looking for plastic Christians, pretending to be perfect. People are tired of pretense. So, why do we often feign perfection before God and each other?

It is hard to let our guard down.
It is hard to be exposed as flawed and imperfect.
It is hard for a wound to heal if we keep it bandaged forever.

It is hard to communicate with others while wearing a mask. We waste so much time and energy keeping the mask in place. At times we wear so many masks that *we* do not know who we really are.

We deprive ourselves of real soul-sharing with others, for no one ever tells their deepest feelings to a mask! We deprive ourselves of meaningful communication with God because our masks shut Him out. God knows and loves us unconditionally, despite all our failings. We never need to try to wear a mask in His presence.

Please take off your mask. I will take off mine, and the journey will be brighter for both of us.

A Different Drummer

Peer pressure is one of the strongest forces at work in our world today. One form of peer pressure we all know too well is our preoccupation with brand name labels:

From the four-year-old boy in his Izod shirt to the eight-year-old girl in her Nike shoes;
From the teenager with her designer jeans to the young man with his starter jacket;
From Mom with her Dooney and Burke bag to Dad with his Brooks Brothers suit. . . .

We are also greatly influenced at all ages by peer pressures in our ethical, moral, and spiritual values and behaviors:

From the push of one child to another to steal that first
  candy bar;
From the taunting dare of one young person to another to
  smoke that first joint;
From the pressure that comes from Dad's friends to cheat a
  little on his expense account or even on his wife;
From the subtle suggestions from some of Mom's friends to
  "tell all" the latest gossip about her best friend.

We are all inclined to give in to our peers in order to be liked, to be accepted, to be popular. But what is popular is not always right.

The apostle Paul had some very strong things to say about peer pressure when he wrote: "Do not be conformed to this world, but be transformed by the renewing of your minds, so that you may discern what is the will of God—what is good and acceptable and perfect" (Rom. 12:2 NRSV).

Or, in the words of Henry David Thoreau, another independent thinker:

If a man does not keep pace with his companions, perhaps
  it is because he hears a different drummer.

Let him step to the music he hears . . . however measured or
  far away.

Do not follow where a beaten path may lead. Go instead where there is no path and blaze a new trail!

Trust

She looks up and her large blue eyes are full of complete trust as she reaches for my hand. I take the small one she offers in my weathered one. We cross the busy street in silence.

We enter the large store which is filled with unfamiliar things to her young eyes. Occasionally she tightens her grip or rewinds her fingers through mine as she seeks a sense of renewed security. It is only when we reach the toy department that she loosens her tight clasp and pulls me by one finger to the counter filled with beautiful dolls.

She is my grandchild. She has placed her hand in mine with complete trust. Oh, God, I am your child. Why do I waste so much time and energy when I forget to reach for your hand of security? Why do I forget your promise to safely guide me through the unknown, around life's detours, and through valleys of uncertainty to seasons of joy?

God, I am your child. Here is my hand. Please take it and help me remember to trust.

Lists

Many of us live by lists. Sometimes, our lives are ordered and controlled by the lists we make. Some hectic days the humorous thought on a little plaque sums it up for me: "God put me here to accomplish certain things—right now I'm so far behind, I don't think I'll have time to die."

Each autumn my husband and I like to go to New England to soak up the glorious foliage colors in that corner of God's world. I find myself making lists today for that special trip. There is the list of things we must do before we leave:

- stop the mail;
- cancel the newspaper;
- take a house key to a friend who will water our plants;
- prepare and mail itineraries to family and neighbors;
- make arrangements for someone to mow the lawn;
- shut off the water and turn down the water heater;
- set the timers on lamps. . . .

There is a list of the things we want to be sure to take:

- maps and motel directories;
- warm fall clothing;
- camera and plenty of film;
- a New Testament and other books to read;
- address book;
- car phone;
- date book (No, I'll leave that at home.). . . .

And now, as I look on my desk, there is another growing list of things I must remember to do as soon as we get home. . . . WHEW!

When these lists have all been checked off the morning we leave, I always wonder, "Is this trip really worth all of this preparation?"

As I make and review all of these lists, I am reminded of God's Great List, the Ten Commandments, by which we are to live. Another challenging list was drawn up by the late Mahatma Gandhi, a great religious leader in another part of the world. He called them the seven deadliest sins:

- wealth without work;
- pleasure without conscience;
- knowledge without character;
- commerce without morality;
- science without humanity;
- worship without sacrifice;
- politics without principle.

Another even more important list was given to us by Jesus Christ who condensed the negative "Thou shalt not" commandments of the Old Testament into a short list of just two positive commandments when He said:

"You shall love the Lord your God with all your heart, and with all your soul, and with all your mind. This is the great and first commandment. And a second is like it, You shall love your neighbor as yourself."
—Matthew 22:37–39 (RSV)

I believe our lives will be abundant when we follow Christ and His Short List! That's good enough for me!

When God Says No

Our daughter was sixteen years old when we brought her home from the hospital. It was the day before Thanksgiving and our hearts were filled with gratitude and a fervent hope that the last surgery on her left eye would be successful. We had prayed and many others had prayed. We were optimistic that she would keep her eyesight.

But the days and weeks that followed brought a return to the hospital with more pain and several more surgeries.

We continued to pray for her recovery, many others continued to pray, but again and again the answer to our prayers seemed to be God's eloquent, "No." Finally, the removal of her eye was all that brought relief.

I thought I had learned to understand and accept God's "No" answers since this was not the first time I had received them. For years I had prayed for improvement or removal of a personal physical handicap. The answer had come again and again, "No." This I had learned to accept.

But somehow this crisis seemed different, and I found myself having to learn the lesson all over again. And so, in these ensuing years, I have learned over and over again:

> that in all experiences God's sustaining love undergirds, enfolds, and embraces us;
> that nothing can separate us from the love of God;
> that the bonds of family love and devotion are more tightly drawn together during a crisis;
> that my faith was strengthened and deepened as I sensed her faith;
> that the important thing is not what life brings to us, but the way we react to what it brings.

Someone has said, "If fate throws a dagger at you, there are two ways to take hold of it—by the blade or by the handle." You may catch it by the blade and allow it to cut you, wound you, and make you bitter. Or, you may catch it by the handle and use it as a means of winning new victories.

Each new crisis demands new understanding and acceptance. Perhaps it is because of my discovery of this that I am sharing this account with you in such a personal way. It is my hope that through my sharing you may store up strength in your heart, against that day when God's eloquent "No" rings out in your life.

Use What Is Left

Everyone of us has known some kind of loss from time to time throughout our lives: the loss of a job; the loss of a parent; the loss of a home; the loss of a spouse; the loss of health; the loss of our independence when our bodily mobility is threatened. . . .

During a time in my life when I was dealing with a deep sense of loss, there was a brief quotation on my bedside table which I read over and over again:

> Not everyone can find healing for his physical condition, but everyone can change his attitude toward his condition. *It is not what you have lost but what you have left that counts.*
>
> —Harold Russell

I have always been fascinated by the Shakers, a unique Christian community which thrived in the 1880s. Several years ago my husband and I visited the Shaker center at Canterbury, New Hampshire. Two aging eldresses were still living there at the time of our visit.

A few years prior to our visit, Eldress Bertha had completely lost her eyesight. Later, a friend asked her if the darkness was not overwhelming now that she could no longer see the surroundings she had always loved. "O my, yes, I suppose it would be," she replied. Then, as she paused, she smiled as she continued, "It might be if it weren't so bright on the inside."

It was one of those freak accidents but it left the young athlete totally blind. The doctors told him, "You will never see again." The social worker told him, "You must learn braille, stay at home, and accept this. Learn to let others help you and adjust to a lifetime of dependency."

But Morris Frank had a *lot of himself left* and he fought to regain his independence. The result was the development of The Seeing Eye Center in Morristown, New Jersey, where leader dogs are trained to serve as guides for persons who are blind.

In a little town in the French Pyrenees, there is a shrine noted for miracles of healing. Shortly after World War II, an amputee appeared at the shrine. As he hobbled up to the shrine on his crutches, someone commented, "That poor man, does he think God will give him back his leg?"

The veteran, overhearing the question, turned around and, looking the stranger straight in the eye, retorted, "Of course I don't expect God to give me back my leg. I am going to pray that God will help me live without it."

Johnny Unitas was one of the all-time great football quarterbacks. On one occasion he wrote: "No matter what your limitations may be, you have to look at what you *can* do, not what you *can't* do. And whatever you can do, go out and do it well. If you're breathing, you're living."

Jane Merchant was born in 1917 into a Christian family of struggling East Tennessee farmers. It was soon learned that she had an incurable brittle bone disease. At one point, when her mother was lifting her from her wheelchair to her bed, bones were broken in both an arm and a leg. She was confined to her bed from the time she was twelve years old. Books and a window above her bed became her windows to the outside world.

By the time she was twenty-three years old, she had become almost completely deaf and was nearly blind. She lived until 1972 when she died at the age of fifty-three years.

Life was not fair to Jane Merchant. She lost many of the things you and I think are essential for a happy and productive life. But she developed and used what she had left—a wonderful mind and a unique God-given talent for writing.

Upon her death, the Knoxville, Tennessee, *Sentinel* wrote of her, "Jane Merchant was a great poet and a great woman with indomitable courage."

Her biographer said, "She seemed to be one of the most alive humans I have ever encountered."

She used what she had left.

Stitches of Love

Stitches of Love

*F*t is that time of year again. I have taken the treasured quilts from the old blanket chest, lovingly smoothed out the folds, and put them on each bed for a light summer covering.

This annual ritual brings me joy as I remember my Grandma Martin. She was a quilt maker *par excellence*. She did not have one of the smaller portable quilt frames that quilters use today. Her quilt frame was the large, unwieldy kind that filled half her dining room when she was quilting.

Piecing and quilting quilts was one of her favorite hobbies. She was a master of design and choice of color. Unlike the radiantly deep colors of the lovely Amish quilts, grandma preferred the delicacy of pastel colors—soft pinks, pale yellows, and cool greens to go along with the prints that she used.

A few of her quilts were pieced from remnants contributed by her neighbors and friends or a pretty print from an emptied chicken feed sack or a design saved from one of her favorite house dresses.

But most of the quilts she made were created from intricate patterns and new fabrics she ordered from a favorite mail-order catalog which offered supplies to quilters.

She made quilts in many of the popular patterns of the day— the flower garden, the double wedding ring, the Philadelphia pavement, the Texas star. . . . But she also made several "one of a kind" applique quilts with intricate designs.

Of all those she made, I think my favorite is the one she gave me on my graduation from high school. It is a postage stamp quilt. Each piece in the quilt is the size of a postage stamp. The tiny pieces are sewn together to form brown baskets filled with multi-colored flowers.

I recall that the ends of grandma's fingers were frequently covered with adhesive tape. She would quilt until the ends of her fingers were pricked open and bleeding. Then, out came the tape to give her added protection so she could continue quilting. Her quilts were surely labors of love for her family.

How like a quilt is your life and mine. Many bits and pieces go into the making of each quilt. Many persons and experiences come into our lives which help to form and shape us. And somehow, if we are willing, God takes all of these pieces that we give to Him and He shapes them into a beautiful design with His stitches of love and caring. Then, He gives them back to us to use as we love and care for one another.

Be Still and Know

F have had a good morning. I have spent most of it at my ironing board. I have just ironed a linen tablecloth and the matching napkins which I used recently. I do not use them much anymore since I now have some pretty no-iron linens.

When I was growing up in my parental home, one of my favorite household tasks was doing the weekly family ironing. I especially loved to iron my father's shirts. My mother always seemed more than happy to let me do them. I realize now that she probably rejoiced when I did them. My wise father would comment, "Mary, I can always tell when you have ironed my shirts."

Later, my love of ironing was transferred to my little girl's frilly, fluffy ruffles. I still enjoy ironing. With the return of more cotton garments, my ironing hours have increased.

The ironing process requires no special thought concentration for I can smooth out the wrinkles out of habit. So, it offers moments when I can push my thoughts aside from my work and invite God to come into my thinking in a very special way. Ironing affords me time to become a better listener to God. We never can really know His plans for our lives unless we make and take time to just listen to Him.

I always put a pencil and paper on the countertop next to my ironing board when I get ready to iron. Sometimes, thoughts rush into my mind and I feel the need to write them down. Many of my meditations are the products of my "ironing board hours."

On other days, because my mind has been crowded with a number of thoughts about many things, I have been less receptive. When that is the case, no fresh thoughts come to me. On those days, when the ironing is done, I feel only physical fatigue.

In the hectic, hurried hours of our days, we should snatch and claim every moment. Someone has reminded us that "we must be still if we would know, for God will not stoop to shout." The still, small voice still speaks, but many times it just whispers.

A Gift of the Past

An antique pressed-glass compote filled with fruit sits in the middle of my kitchen table. It belonged to my Great-Grandmother Kersey. I remember that it always sat in the middle of her table and was filled with fruit.

Across the years my efforts to identify the glass pattern have met with failure. Perhaps the piece does not have a pattern name. It may not even have much monetary value. The real value of this piece of glass is found in the wealth of memories that flood over me as I use it.

Just now, as I sit here admiring it, I am reminded of Paul's words to his beloved spiritual son, Timothy:

> I have been reminded of your sincere faith, which first lived in your grandmother Lois and in your mother Eunice, and, I am persuaded, now lives in you also.
> —2 Timothy 1:5 (NIV)

What a marvelous Christian heritage Timothy had! Not only did he have a Christian mother and the benefits of her teaching and living, but he had a Christian grandmother as well.

And then, Paul affirms this unfeigned faith that was at work in Timothy's life. Webster defines *unfeigned faith* as "genuine sincere faith." In current day terms, we would say Timothy's faith was authentic—it was real.

Not only did I have a Christian mother and a Christian grandmother, but I also had a Christian great-grandmother.

With privilege comes responsibility. For those of us who are blessed with the heritage of deep Christian nurturing comes the responsibility to live out a vital Christian faith in our lives everyday.

Today we are hearing so much about the importance of family values. We should be very conscious of the Christian example we are passing on to our children, grandchildren, and to the children about us in our neighborhoods and churches. Today I want to pay tribute to my Christian Great-Grandmother Kersey who lived until I was eighteen years old. Her pressed-glass compote will adorn my kitchen table for years to come.

Lilacs

One beautiful spring morning a garden club friend brought me an armful of fragrant lilacs. The next morning when I awoke, the fragrance of those lilacs had permeated the entire house.

The words of my friend as she handed the lilacs to me came back. She had said that some of the lilacs were the "old-fashioned kind" and some of them were a new variety of French hybrids.

As I recalled her explanation about the lilacs, the thought came to me that perhaps that is the way our lives should be.

I have just read my morning devotions from our time-worn Bible as I cooked my instant oatmeal for one and one-half minutes in our microwave oven.

I still beat my angel food cakes with an old wire whisk in a heavy earthen crock, but then frost them from a can of Betty Crocker's frosting.

We like to grind our coffee beans in an antique coffee mill, but brew them in our latest automatic drip coffee maker.

Our home is completely furnished with antique furniture, yet I wash our clothes in an automatic washer and dry them in an electric clothes dryer.

Yes, our lives are full of things of the past and things of the present. Someone has suggested that the things of the past connect us with one another, whether we are sowing seeds, rediscovering a favorite story to read to our grandchildren, or using an old favorite recipe that was given to us by our grandmother. Our homes and our gardens are perfect places to preserve these treasures of the past for the next generation.

You and I are responsible for sharing with the next generation those things that are beautiful, steadfast, honorable, and kind—and all of those things that sustain us now, even as we stand on tiptoe to catch the newness of each tomorrow.

Yes, I am thankful for the fragrance of the old-fashioned lilacs, but I am also thankful for the beautiful deep color and shape of the blossoms on the new French hybrids. They are God's blend of the old with the new.

Red Geraniums

Red geraniums have always held a special meaning for me. I remember the red geraniums which sat on the sill of a south window in my great-grandmother's Iowa home. When the "January thaws" came to Iowa and warm spring days followed, she planted them outside. They were bright blotches of brilliant red in her summer flower beds. Before the first killing frost in the fall, she brought them back into the house. By the time there was snow on the ground, the geraniums were giving off their red radiance again, filling the room with cheer.

Many winter days are dark in Iowa when the heavy snow clouds hang low. On many of those dark days she would say to me, "When it is dark on the outside, we have to make our own sunshine on the inside."

My great-grandmother was a sunshine maker! She practiced that philosophy everyday that I knew her. Each day she stored up a simple faith even as our modern solar panels store up the energy of the sun for use later. Her well-worn Bible was her daily companion. I still recall the sincere prayers she prayed over me. She surrounded herself with simple tasks for her family and friends. She patiently took the time for labors of love. Her molasses cookies were the greatest!

You and I live in a culture of instant puddings, instant weight loss, and even instant job success. However, experience teaches us that instant pudding is full of air, instant weight loss quickly returns, and instant success is often short-lived.

Just as grandmother carefully prepared the soil and regularly watered those geraniums, when they were both inside and outside, so we must take the time to nurture a real faith that will sustain us in our dark days. We must be our own sunshine makers if we would have bright red geraniums on our winter windowsill.

Celebrate!

A Clean New Year

At the end of every year the newspaper, radio, and television news media remind us of the persons and events which have highlighted the news of the closing year.

While it is often true that life can only be *understood* by looking backwards, we know that it must always be *lived* by looking forward.

As I put these thoughts on paper, a soft, gentle snow has been quietly covering the ground outside my window. As yet, no footprint or car track has broken its pure white, untouched beauty. It reminds me of the untarnished freshness which the New Year offers to each one of us.

Some months ago, we made our first visit to the Canadian Maritime provinces—New Brunswick, Nova Scotia, and Prince Edward Island.

In anticipation of our visit to Prince Edward Island, I reread a familiar childhood novel, *Anne of Green Gables*, by Lucy Maud Montgomery. The story was set on that small agricultural island.

The author created a wonderful nymph-like character in Anne. Anne was an unpredictable and very loving orphan girl who lived life on the very edge of every day.

One of my favorite lines in the book is Anne saying to Marilla, her guardian, "Isn't it nice to think that tomorrow is a new day with no mistakes in it yet?"

We might paraphrase her words to say: "Isn't it nice to think that the New Year offers to each of us 365 new days with no mistakes in them yet?"

The New Year offers us many days
to walk closer with our Lord,
to think higher, nobler thoughts,
to be kinder in all our relationships with those around us.

Then, both the being and the doing of our lives everyday will bring joy to us and to those whom life has placed around us.

I asked the New Year for some message sweet,
Some rule of life with which to guide my feet.
I asked and paused:
He answered soft and low,
"God's will to know."

—Author Unknown

God bless thy New Year!

Throw It Away and Fix It Up

We are all aware that this is the season of the year when we feel motivated to sort out, to throw away, to clean up, and to fix up. We do this one drawer, one cupboard, and later, one flower bed at a time.

After the sorting out and discarding process, we are ready to add a new coat of paint to a room, hang bright new curtains to sparkling clean windows, or later plant new seeds in the flower bed and garden.

God started this whole process of renewal when He first showered His earth with spring rains and warmed it with bright sunshine. Our cold gray days of winter will take on new beauty and color—greens, yellows, purples, and pinks—as fresh new leaves and spring flowers burst forth.

Likewise, He has given us an important time to sort out and throw away, to clean up and fix up our lives so that we, too, can take on new life. We have begun a new season of our soul we know as Lent.

Some years ago, Maxie Dunnam wrote a little booklet for our Lenten reading. One meditation was about sorting out and throwing away:

> We can discipline our wills by practicing with little things. We can check the desire for an extra unneeded helping of food; resist the temptation to gossip, or even refrain from talking too much; nor should we ask the question that would unnecessarily pry into the life of another.

There are things in each of our lives to sort out and throw away forever. This is, however, only part of Lent. This is the discipline of getting rid of, of doing without, of denying ourselves.

There is also an upbeat positive "fixing up" that will take place in our lives if Lent is to be complete. Here are some disciplines we might add:

> We will set aside more time in our daily devotions when we study the life of our Lord and listen for His directions to us through prayer.
> We will add a weekly call to a friend to our routine.
> We will begin to read that challenging book we have been waiting to start.
> We will add an encouraging smile to our countenance.
> We will write to a friend who is hurting.

Lent offers us forty days to clean those negative things from the cobwebby corners of our hearts. At the same time, we are challenged to add some bright new positive things. And may those good things linger in our lives *long after Lent*.

Today Is Yours to Live

After one of her visits, I found three heart-shaped notes which my fifteen-year-old granddaughter had left for me. One was on my dressing table, another was in my jewelry box, and a third one was tucked into my lingerie drawer.

One read, "Granny, I love you. Amy." The second one read, "Have a great day! Amy." The third one read, "Dear Granny, live each new day to the fullest and always remember this verse, Philippians 4:13." When I opened my Good News Bible to look up the verse, it read: "I have the strength to face all conditions by the power that Christ gives me."

Her notes said two things to me:

1. No matter what this new year brings to me of life's uncertainties—life's "gray times"—if my relationship to Jesus Christ is strong and if I have the support of family and friends, I can be happy.

2. We are each given only one day at a time to use as we wish.

In a recent issue of *Victoria* magazine, a writer quotes these words by Pierre de Rousard:

Wait not till tomorrow,
Gather the roses of life today.

Dr. Ira McBride is the father of one of my college friends. He was a missionary in Africa for many years and is now more than ninety years old. In his Christmas letter he wrote, "I look forward to every new day!"

An *Upper Room* meditation by Barry Stater-West challenges us to "cradle each new day in our arms with love and behold it as a precious gift from God."

This day brings us opportunities to witness and to serve those around us. This day can bring us a special moment to help another. Let us be sensitive to what this day can bring.

The hymnist, F. R. Havergal, put it so well in this verse from a New Year's hymn:

Another year is dawning.
Dear Father, let it be
In working or in waiting
Another year with Thee.

Beginnings

*M*y heart is overflowing with joy, real joy this morning. That spark from God which is down deep within me wells up in response to the beauty outside my kitchen window.

My gaze rests on a small cloud-like formation cradled in the arms of the dogwood trees that have been planted on our hillside by the Master Gardener Himself.

As I revel in their beauty, the age-old legend of the dogwood flashes across my mind. If the story could be true, what delicate beauty God uses to remind us of that cruel Friday so long ago.

As I touch the blossoms, I notice that each white petal is of soft texture and has a brown indented stain. Nails do rust with time.

And then, I see the green center, and the crown of thorns already lifting up tiny tendrils as they wait to be pollinated by the bees or the wind in order for life to start all over again. They begin as:

- a tiny shoot; then
- a tender sapling; and then
- a twisted and gnarled tree.

The new blossoms will be just as beautiful as those I see before me. For these too, will be like the wild dogwoods which God has planted among the other trees in our woods.

My joy stems from the promise and hope of new beginnings in our lives over and over again. God continues to plant the seeds of those new beginnings in hearts everywhere.

Is it any wonder that the dogwood blossoms remind me that every day can be a new beginning?

Easter People

I know that my Redeemer lives:
What joy the blest assurance gives!
He lives, he lives, who once was dead;
He lives my everlasting head!

Many of us have sung the words of this beautiful hymn. The hymn affirms that because Christ lived, died, and arose, you and I may have eternal life.

My understanding of eternal life as a child was simple and limited. I believed that if I accepted Christ as my personal Savior and tried to live the Christian life, I could claim this promise of eternal life in heaven when I died.

I recall so vividly when I was about eleven years old, this concept was changed and redefined for me. A minister, speaking at a youth rally I attended, explained that life eternal does not begin *after* we leave this earth, but *now* while we are living on this earth. If, then, eternal life begins here and now, I realized that the quality of this life depended on one basic thing—my relationship to Jesus Christ here and now.

Two words we use interchangeably many times lift up for me the positive meaning of eternal life. Webster defines *immortality* as "an unending existence"—a condition of simply existing forever. *Eternal life*, on the other hand, means "a positive quality of life beginning now and lasting forever."

Surely, life with purpose and meaning, not merely existence, is what Christ meant when He talked about the abundant life. Our belief in the resurrection of Christ and His promise of eternal life assures us of the power for upbeat daily living here and now.

Sir Wilfred Grenfell, the physician-missionary who spent his life in service to the fishermen of Labrador wrote:

The conscious possession of an eternal quality of life lifts men and women above the sordid and commonplace. Eternal life means that even now, not in some far-distant heaven, we are beginning to share in the goodness and the truth and the beauty of God's creative life. This is His promise; not of a negative existence, but a positive dynamic growth and development of the God-given powers that are already latent within us.

The risen Christ has promised that He will guide us, strengthen us, comfort us, and lead us into the next stage of our eternal life.

Because of Easter, we each have a choice to make:
Eternal life or empty existence.

Oh, God, help us to live each day as triumphant Easter people! Alleluia! Alleluia! Amen!

Take Time to Hear the Angel's Song

*T*his morning I awoke with a feeling of tension and stress in my whole body. After a moment or two of wakefulness, I realized that my last thought as I dropped off to sleep had been about how much there was to be done before Christmas.

During the week after Thanksgiving we get caught in the wedges of stress and strain which bind us captive throughout this entire holy season. By Christmas Day we are victims of inner and outer exhaustion.

Our daily lives are programmed by the clock and the calendar. We are under the constant pressure to produce, especially during this season of the year when the clock seems to tick faster and the days take wings and fly beyond our reach.

In addition to our normal routine there is:

- shopping and wrapping
- addressing and writing
- polishing and decorating
- sewing and fitting
- practicing and performing
- baking and delivering
- cooking and entertaining. . . .

We unknowingly bring many of our pressures on ourselves. This stress may sweep over us because of our poor planning, attempting to do too many things, or making plans that are too elaborate.

But an old Greek motto reminds us: "You will break the bow if you keep it always bent."

Someone else has humorously given us two cardinal rules for dealing with our stress:

1. Don't sweat the small stuff.

2. Everything is small stuff.

During this season it is good to remember that the simple things done with gentle care may be our greatest gifts to each other. Let us take time to hear the angel's song.

God used a moment, a mother, a manger, to give the greatest gift the world has ever known. . . .

Dear Father, Help me to attend to my little errands of love early this year so that the brief days before Christmas may be unhampered and clear of the fever of hurry; save me from the breathless rushing that I have known in the past. Grant me a calm serenity in my soul. Amen.

Angels

In recent years there has been a growing interest in the phenomenon of angels. Billy Graham has devoted a complete book on the subject. At Christmastime, the youth of our local church studied about angels. Some of the topics were: "Have we heard angels on high?" "Are angels in our midst?" and "Do we believe in angels?"

The word "angels" is used in the Bible to translate both Hebrew and Greek terms that are used to identify messengers. In early popular Hebrew thought there was also a belief that a rich background of supernatural beings were attendant to God. They have been called God's ministers, God's servants. Popular beliefs are seen in the New Testament writings about countless hosts of angels. The Christian Church has adapted the belief in angels and used it to express the experience of the unseen spirit world of guardian angels who are sent by God to minister to us.

My daughter told me of a song by Michael W. Smith which speaks of angels. The song is entitled "Emily" and the lyrics repeat, "You're an angel waiting for your wings."

We are now in the Advent season—a time of waiting for the coming of God incarnate into our lives. There are at least two special ways that God comes into our lives: He comes through events, and He comes through persons. He sent Jesus Christ into our lives to change them. Almost everyday He sends persons into our lives to enrich them.

Think for a moment with me about those special persons God has sent into your life this past year. They were persons who entered into your life and gave of themselves to you in special ways:

> Perhaps it was a close family member—a husband or a child whose counsel and love sustained you;
> Perhaps it was a doctor who enhanced the quality of your life through his knowledge and healing hands;
> Perhaps it was a friend who celebrated a special joy with you;
> Perhaps it was a minister or a Sunday School teacher whose teaching helped you to grow in your faith journey;
> Perhaps it was a neighbor who came when you were grieving over the loss of a loved one; or
> Perhaps it may have even been one who was so "down-and-out" that as you helped him or her, he or she also ministered to you.

God does minister to us through other persons. That is the way He has chosen—His only way. Perhaps you and I know some angels today to whom we would say, "You *are* an angel just waiting for your wings." Let us give thanks to God for them.

Faces at the Manger

The New Testament does not give us many details about the birth of Jesus. In fact, only two Gospel writers—Matthew and Luke—even wrote of His birth. Matthew told of the visit of the shepherds and Luke reported the visit of the wise men. In our minds we have put both accounts together to make a more composite Christmas story.

But surely, my imagination tells me that there were other faces around the manger during that two-week period when Mary, Joseph, and Jesus were there.

Bethlehem was only a village, so the inn was probably not a large building. One translation reads, "There was no room in the *house.*" It may have been just a "bed-and-breakfast." We do know that it was completely full. The innkeeper offered Joseph the very best he had. That is all God ever expects of us!

The face of the innkeeper was surely at the manger as he went to the stable to check on Mary and Joseph. I have a feeling that when the innkeeper saw how young Mary was, really just a teenager struggling in her labor, his face must have been one of concern.

I think he might have gone back to the inn and sent his wife out to help. She may have had experience as a midwife in the village or perhaps in their own inn. As she worked over Mary, hers was surely a face of compassion as one woman who understands another's travail.

Surely, there were the faces of young stable boys who cared for all the animals. Not only were there the animals belonging to the innkeeper, but the travelers staying in the inn that night had sheltered their animals in the stable also. I imagine that these boys were shocked to see people being quartered with the animals. They probably had never witnessed a human birth before. Theirs must have been faces of amazement.

We know that the shepherds' faces were around the manger. They were rugged outdoor workers. The manger scene was an earthly symbol for them. It was in the cave filled with the sounds and smells of animals that they found Jesus. Through the manger we sense God's attempt to help us understand that He cares about the common people, yes, even the street people. He is concerned about each of us even as we go about the mundane day-to-day routines of our lives. So, is it any wonder that the shepherds' faces were full of awe and wonder?

Twelve days later there were the faces of the wise men at the manger. Their faces must have shown understanding and satisfaction from knowing that at last they had found the long-awaited King.

One of my favorite modern Christmas carols tells of the little drummer boy whose face might have been seen around the manger. Here are some of the words of that carol:

> Little baby, I am a poor boy, too.
> I have no gift to bring
> That's fit for a king.
> Shall I play for you on my drum?
>
> Mary nodded, the ox and the lamb kept time.
> I played my drum for him.
> I played my best for him.
> Then he smiled at me.
> Me and my drum.

The face of the little drummer boy would surely have been aglow with simple love and faith.

What will the Christ child see in our faces as we kneel before His manger once again this year to celebrate His birth?

The Greatest Gift

In the sixth month the angel Gabriel was sent from God to a town in Galilee called Nazareth, with a message for a girl betrothed to a man named Joseph, a descendant of David; the girl's name was Mary. The angel went in and said to her, "Greetings, most favoured one! The Lord is with you." But she was deeply troubled by what he said and wondered what this greeting might mean. Then the angel said to her, "Do not be afraid, Mary, for God has been gracious to you; you shall conceive and bear a son, and you shall give him the name Jesus."

—Luke 1:26–31 (NEB)

God surely chose the most unlikely person to bear His Son. Mary was a poor, young peasant girl. She was engaged to Joseph and looked forward to being his wife. She was a devout Jew and was as familiar with Jewish law and custom as a woman of her day was allowed to be.

What her thoughts must have been when Gabriel appeared to her and told her she was to become pregnant with God's Son, who would be the Messiah. There had to be many moments, even hours, of doubts and anguish for her as she tried to understand what was happening. I am amazed by her obedience to God's plan. She gave unconditionally her healthy, young body to carry the child. With her gentle, serene spirit she prepared herself emotionally for Jesus' birth. She gave God her greatest gift—the gift of herself.

God has created each one of us so differently—surely, one of the greatest miracles of His creative activity. He has given each of us gifts to give back to Him—even as Mary gave with unselfish abandonment:

- our gentleness;
- our kindness;
- our caring;
- our sharing;
- our listening;
- our loving. . . .

He can take the gifts we give Him and multiply them a hundred fold, and give them back again to this hungry, weary, and struggling world.

J. C. Penney once wrote, "The Yuletide season must be an occasion for giving infinitely more than material things. It must be a time of giving oneself." This is the season of our gift giving. Let us reexamine the kinds of gifts we give this year.

In the middle of the nineteenth century, (1830–1894), Christina Rossetti penned these words:

What can I give Him, poor as I am?
If I were a shepherd, I would give him a lamb.
If I were a wise man, I would do my part—
Oh, what can I give Him?
I'll give Him my heart!

Our lives are like the basket with the two fish and five loaves. They never seem to be enough until we start giving them away.

A Golden Gift

Give a golden gift this Christmas,
A gift of meaning and worth
For the art of giving unlimited
Began with God's gift to earth.

What can we give that is precious?
A gift of beauty and grace.
A gift of cheer, a gift of love,
A gift to bring joy to a face.

Let us give of ourselves this Christmas,
'Tis the most precious thing we possess.
The gift of a smile, a word, a loaf
Surely, we can give no less.

Give this kind of golden gift this holiday
God's gift of Himself shows us the way.

Recently, a letter came in the mail from my sister. She asked me to give her ideas for Christmas gifts for us. As I thought about her request, I realized there was absolutely nothing that we really needed, and very little we even wanted.

Her request prompted me to begin work on our own Christmas gift list. I have had real difficulty thinking of things to buy for each person on our list, too.

I realize that we all live in a "thing-oriented" society which has invaded and overshadowed the real spirit of Christmas giving.

I heard about a little child who wanted to give her grandmother a very special Christmas gift. She asked her grandmother what she would like for Christmas. The grandmother replied, "Just give me a kiss for Christmas. I don't have to dust that!" How cluttered our houses and our lives have become with "things."

Perhaps we might plan an alternative kind of gift giving this year. Try giving the most precious thing you have to give—the gift of yourself, a golden gift:

- a long distance telephone visit;
- a special letter of appreciation;
- a cash gift to honor a friend and feed a hungry child;
- the promise of a ride into the country for one who is homebound;
- an invitation to someone who is alone to stop by your fireside for a cup of tea.

The Manger and the Star

There is so much pageantry at this season that is difficult for us to slash through the tinsel and ribbon to the real meaning of Christmas.

G. Ray Jordan has insisted that Christmas is not primarily a day or a season. It is essentially an experience. It is intended to remind us how much God loves us. He came to earth in Christ in order that we might be saved from our sin and know of His everlasting love for us.

There are two very simple, yet profound truths in the Christmas story which are symbolized by the manger and the star.

The manger was surely an earthly symbol. The Bethlehem cave was full of the smells and sounds of animals. Mary gave birth on a bed of scratchy straw. We can imagine that Joseph had emptied the feed from one of the mangers and had filled it with fresh straw. It was here they laid Jesus.

In the manger we sense God's attempt to help us understand that He cares for us. He is concerned about us in the mundane, daily, routine experiences of life. It is a symbol of God's down-to-earthness.

Of all the fascinating stories of the Bible, none is quite as thrilling as the story of three wise men who followed the star from somewhere in the east. We are not sure who they were. One theory tells us that one of them came from Asia, another came from Europe, and the third one came from Africa. They were obviously men of wealth and occupied positions of prestige and influence. The star and the wise men are symbols of the mystical dimensions of the Christmas event.

It is a fantastic thought to realize that the God of the universe who created and sustained it actually came to live on this planet. In Him we become aware of those divine qualities of life which speak to our deepest needs.

When the first spark of desire after God arises in thy soul, cherish it with all thy care, give all thy heart unto it. Follow it gladly as the wise men of the east followed the star from heaven. It will do for thee as it did for them.

—William Law

The star that rose at Bethlehem has never set.
It glows for them who seek its light! 'Tis leading yet.

They saw the star and they alone who longed for it.
For men like them the star that shone in Bethlehem will never set.

—J. C. McCoy

And the miracle of Christmas for you wherever you are is God's blend in your life of the manger and the star.

About the Author

Mary Jane Hartman has an uncommon sensitivity to the beauty and spiritual messages of the ordinary. Her appreciation for life's gifts led her into a career of music and teaching, graduating from Westmar University in Le Mars, Iowa, with a degree in music education in 1945. Active in the Belle Meade and Brentwood United Methodist Churches in Tennessee, Hartman has locally been acknowledged as United Methodist Woman of the Year in 1996 and has served as a conference devotional leader and as a member of the Women's Division of the Board of Global Ministries. Hartman has previously written devotional meditations for other publications including *Images: Women in Transition*.